PLANT DISEASES AND THEIR IMPACT ON AGRICULTURE

Prof. Iqbal Hasan

RAJAT PUBLICATIONS
New Delhi

Plant Diseases and their Impact on Agriculture
by *Prof. Iqbal Hasan*

This edition is published by Rajat Publications
4740/23, Ansari Road Daryaganj,
New Delhi - 110002

ISBN: 978-81-7880-423-1

Edition 2025

Printed and bound in India

RAJAT
PUBLICATIONS

4740/23, Ansari Road, Daryaganj,
New Delhi- 110002 (India)
Mob:- +91-9811484973, +91-9811105389
E-mail: rajat_publication@yahoo.com

Preface

Plant Disease and their Impact on Agriculture presents comprehensive coverage of the cultivation of an improved cultivar possessing the needed disease and insect resistances. It is important to maintain a production base that allows maximum sustainable yield and optimum use of inputs whether they are fertilizer, water, land or time. Apart from good agronomic management of the crop, seed production differs from grain production in land requirement, isolation and prevention of contamination, and must meet specific quality standards prescribed by the national seed regulations. The private sector can, however, ill afford to focus on wheat seed, due to its characteristic low-profitability.

Furthermore, in many countries there has been no ongoing effort to promote the use of improved seed by wheat farmers, and no significant breeding developments have recently taken place. It is, therefore, expected that in the future the large majority of resource-poor, small-scale farmers in many developing countries will have to rely on seed saved from the previous harvest. Control of insect pests of wheat is often secondary to the more pressing problems of diseases and abiotic constraints, such as heat, drought and salinity, and thus the development of cultivars possessing resistance to insects has for the most part lagged behind the development of high-yielding, disease-resistant cultivars.

Cultural control techniques are more common since they usually require little new farmer input. This book has been formatted to cover several aspects of plant disease to cater to the need of students at graduate and post-graduate level. Several courses are offered at graduate level and it is difficult to have separate books for each course. Therefore, a broad treatment of the subject has been given here to cover wide-range of topics.

Prof. Iqbal Hasan

Contents

Chapter 1

Agricultural Impact of Vegetables Diseases

TOMATO

LEAF MOULD

Infection occurs mainly on the leaves, occasionally at the stalk end of the fruit. Leaf mould always appears first on the older mature leaves. The fungus enters through the stomata and the mycelium penetrates first between the leaf cells, afterwards into them. The dark conidiophores emerge in bunches through the stomata on the under-side of the leaf and bear numerous tiny elliptical conidia, mostly two-celled. They form a greyish bloom on the spots that turns a purplish brown with age.

No conidia are produced on the upper surface of the leaf but yellow blotches appear corresponding to the mould spots on the under-side. Gradually as the disease progresses the leaves turn yellow, curl and finally shrivel completely, thus greatly weakening the plants. It takes the form of a black discolouration spreading outwards over the fruit from the end of the stalk but never covering the whole surface.

The fruit may become lopsided and furrowed as a result of this attack. The fungus cannot attack the fruit directly but penetrates the stomata on the sepals or on the stalk and spreads through these into the young fruit or may prevent the fruit setting.

Seed from infected fruit may carry the fungus and not only give rise to an infected seedling but act as a source of infection for neighbouring plants in the seed boxes. The seed coat containing the fungus is carried above ground on the cotyledons, conidia are produced on it and shed on to the surrounding seedlings.As the fungus overwinters in the soil and on the interior fittings of the greenhouse the house must be disinfected after a tomato crop. This is most easily done by burning sulphur.

One should not pass straight from an infected house into one where the crop is still healthy, nor take tools, baskets, etc., from one to the other. By such means the conidia can easily be introduced into the healthy house.

It is of the utmost importance to prevent excessive atmospheric humidity. Therefore firing should be kept up at night throughout the summer and the house must be kept thoroughly ventilated. It is important to construct tomato houses so as to allow of plenty of side and end ventilation. An outbreak of leaf mould can be stopped altogether by heating and ventilating.

STEM ROT

The disease occurs mainly on the stems of medium-sized and old plants and on the fruit. It may also appear on the stems of quite small plants and sometimes causes dark spots on the leaves but this is exceptional. The rot on the stem forms a dark-brown area with a sharp boundary between it and the healthy tissue.

Dotted over the brown part are,the tiny dark pycnidia. Infections can often be seen to be centred on wounds made by defoliation or other causes. In severe attacks the whole of the plant dies above the infected part. Infection occurs not only through cracks and wounds but also directly through the uninjured skin.

The incubation period is 4 days or more. The fruit is generally attacked only in one place but the rot may spread all over it. The spot usually appears when the fruit begins to turn red. It is then much the same colour as the rest of the skin but has a granular appearance due to the crowded pycnidia.

The latter are usually very close together over the whole surface of the spot and eventually turn black so that the diseased area as a whole looks blackish. Under moist conditions pink globules of conidia ooze out of the pycnidia. Infection at the base of the fruit may cause it to fall.

ROOT ROT, FOOT ROT

The young plants become limp, the stem shrivels at soil level and they fall over. If the damage is caused by a species of Phytophthora or Pythium the shrivelled stem is greyish; if the damage is done by Corticium solani it is rather browner, but it is not always possible to distinguish the two forms of the disease without microscopic examination. Tomatoes are more often attacked by Phytophthora and Pythium than by Corticium.

That of P. cryptogea Pethybr. and Laff. is 75°F and that of Corticium solani is 65°F. None of these fungi causes appreciable damage at temperatures below 50°F. All soil-borne fungus attacks can be prevented by sterilizing the soil by heat. Pythium and Phytophthora may also be controlled by formalin treatment of the soil before the seed is sown.

Watering with Cheshunt compound can also be done to disinfect the soil before sowing and to check an outbreak of Phytophthora damping off in the early stages. Solutions of corrosive sublimate and saltpetre and proprietory preparations containing mercury may be used in the same way against Corticium solani.

Drying the surface soil by raking helps to check Phytophthora but not Corticium. A few outbreaks have been reported but the disease is, no doubt, usually confused with other root rots. Affected plants wilt by day and recover at night. Usually their foliage is darker than normal and the under-side of the leaves often has a purplish tint.

The stem is thin and hard and the leaves are small and pointed. When the plant is dug up or knocked out of its pot the main root is found to be dead with a curiously pointed tip. Usually the core of the root is brown and hard. New rootlets are formed higher up, only to be destroyed in their turn, and eventually adventitious roots arise from the stem above soil level. Infection may occur in the seed box but it is seldom detected at this stage as the only symptom is blackening of one or more of the roots.

In pot plants the main root is usually destroyed and the hard appearance of the tops may develop, but often the trouble is masked by profuse development of secondary roots. Usually toe rot is first noticed after the plants have been set in the border and made considerable growth. Then, when conditions favour rapid transpiration, the damaged root system is unable to supply the necessary water and daytime wilting sets in. Cold soils encourage the disease.

FOOT ROT (COLLAR ROT) OF OLDER PLANTS

The plants are attacked at soil level where a dark, sunken, decaying patch appears on the stem. The side roots are usually dead or brown before this occurs. The main root is also often brown and rotten, its soft parts decay and only the woody core remains.

The disease may resemble stem rot but the lesions of that disease are bounded by healthy tissue above and below, whereas those made by Corticium solani start in the root and spread upwards so that they are bounded by healthy tissue at the top only.

If outbreaks have occurred in previous crops the soil should be sterilized by heat or by corrosive sublimate treatment. When older plants are attacked by foot rot they can sometimes be saved by packing fresh soil round the base of the stem.

If possible soil free from the fungus should be used. The stem then puts out new roots above the diseased part. To prevent the fungus from growing into the fresh soil it should be watered occasionally with solution of corrosive sublimate and saltpetre or with a proprietary compound of mercury. Neighbouring plants should also be watered with the same solution.

Damage by this disease occurs particularly during cold springs and to weak plants, hence if the disease appears in a greenhouse the temperature should be raised. Care should be taken to see the plants make the normal rate of growth.

VERTICILLIUM WILT

Plants attacked by Verticillium wilt sometimes become stunted in growth because the stem ceases to elongate, thus causing the upper part of the plant to become rosette-like. At first the plants wilt only during the day and recover their turgidity during the night.

The lower leaves develop light-orange spots, and gradually this discolouration spreads up the plant and the leaves first affected wither. Infected plants may die quickly or may remain in an unthrifty condition for a considerable period.

In the latter case they may ultimately die or may recover from the attack. The vascular bundles of the stem are stained brown. Usually the fruit is not attacked but occasionally the fungus causes a soft rot. Wilting of tomatoes on hot days may also be due to root trouble, including eelworm infestation.

BROWN ROOT ROT

For several years increasing attention has been paid to this disease in indoor tomato crops. Brown or brown-spotted roots appear in association with poorly developed tops. It has not been proved that the brown roots are the cause of the poor development above ground and the low yield of fruit, but circumstantial evidence makes this highly probable. There is often a small amount of browning on the roots even of plants that arc fruiting well.

Brown roots are often seen also on melon, and the complex black root rot of strawberry seems to be a comparable condition. If plants with healthy roots are planted in " sick " soil distinct brown roots may be found after 3-4 weeks. Even earlier microscopic examination discloses spots in which the root cells have yellow thickened walls while their contents are brown and granular.

The boundary between these spots and healthy tissue is

clearly denned. The brown spots or belts begin on the outer surface of the root, and gradually penetrate more deeply until they reach the vascular bundles, along which the brown discolouration spreads. When the roots are dug up one generally finds zones of brown, swollen and cracked cortex on the thicker roots alternating with zones of healthy cortex having the normal white or a light-brown colour.

In more severe attacks the roots die until only a little tuft of weak fibrous roots remains at the base of the stem; the top is very backward, the whole plant thin and lanky and the fruiting very poor. Many observations indicate that appearance of brown root rot is closely bound up with conditions of cultivation.

In experiments at Cheshunt the disease was worst on overmanured, wet and compact soil and it was concluded that great variations in soil moisture should be avoided. Brown root rot may be very prevalent in a new tomato house, and with continued cultivation of tomatoes the brown discolouration may decrease with successive crops until it almost disappears as soil conditions improve. Replacing the soil of an affected greenhouse often helps greatly but the question as to whence this soil should be taken, which previous crop is most favourable to tomatoes, requires closer investigation.

SCLEROTINIA ROT

The plants wither, sometimes the whole plant dies suddenly, in other instances the lower leaves shrivel first. A rotten area, often covered by white mycelium, is to be found on the stem near soil level.

Threads of the white mycelium may grow out into the surrounding soil. Wrapped up by the mycelium or enclosed by the stem cavity from which the pith has disappeared are large sclerotia, at first soft and white but ultimately hard with a black rind.

Green tomatoes picked to ripen off may be destroyed by the same fungus if they come from infected plants and especially if they have come in contact with infected soil, or been placed on a bench where others had decayed before. They are rapidly converted into a rotten mass overgrown by dense white mycelium containing black sclerotia. Sclerotinia rot may be destructive both out of doors and in houses once the soil has become infected and if conditions are moist enough.

BUCKEYE ROT

Symptoms and cause. The two species of Phytophthora which cause foot rot and damping off of tomato seedlings can also attack the unripe fruit. This develops a greyish spot marked with concentric brown rings round the point of entry of the fungus. Intensive work on the physiological races of potato blight has led to the recognition of a race (A) widespread throughout the world and another closely related race (T) which has a particular capacity for infecting tomato leaves.

Other races produce only larger or smaller spots on tomato leaves. In our experiments the yield of fruit has been increased and the number of diseased fruits reduced by spraying outdoor tomatoes with Bordeaux or Burgundy mixture as soon as blight is first seen on the potato crop, repeating the spray every 12 to 20 days thereafter. In experiments good results have been. obtained with copper lime dusts. Where outdoor tomatoes are picked to ripen indoors spraying definitely improves their keeping quality.

FUSARIUM FRUIT ROT

Symptoms. Infection begins at the stylar scar at the blossom end of the fruit as a small, sunken, soft spot that gradually increases in size and turns dark. From its surface break out pink conidial pustules of the Fusarium, often arranged in concentric rings.

Ultimately the entire fruit becomes permeated, turns yellow, shrivels and dries up. The disease occurs on the outdoor crop. The attack is by the above-mentioned species of Fusarium which penetrates the fruit through the tiny wound left by the fall of the style.

It cannot penetrate the unbroken skin but on picked fruit

it easily enters by the stalk scar and may cause serious loss of outdoor tomatoes ripening in store. The fungus needs a high temperature and high atmospheric humidity for active development. The disease almost invariably occurs on the lower trusses of fruit. The fungi concerned inhabit the soil and attack fruit in contact with the ground, or those on to which they have been splashed during watering. Buckeye rot is well known in many countries where tomatoes are grown and has caused great loss from time to time.

VIRUS DISEASES OF TOMATO

COMMON TOMATO MOSAIC

Symptoms vary from ill-defined, light-green spots, mainly between the veins of the leaflets, to quite a bright, clearly defined, light and dark-green mottle. The surface of the leaves may also be slightly uneven or even curled. Tomato variety, nutrition and the time of year all affect the expression of symptoms; thus, during the sunny warm days of summer the mottling is most obvious, and during spring and autumn the curling becomes apparent. At low temperatures the symptoms may be almost entirely lacking but the plants remain infected and infectious to others. Flowers of infected plants may be deformed, sterile and fall off. There may be slight streaking of the stems and pale, slowly ripening spots on the fruit. Infection from tomato plants. The invisible infectious matter, or virus, occurs in the sap of almost all the cells of an infected tomato plant, including the cells of its numerous hairs.

·These are easily broken and it is a familiar experience that one cannot handle a tomato plant without getting sap on one's hands, tools and clothes. If one touches an infected plant and then a healthy one a mere trace of sap transferred from one to the other is enough to transmit the infection. Hands, knives and other tools are the most important agents of spread. Aphides do not spread this disease. The incubation period is from 5 to 20 days or 28 days in our experiments out of doors.

It persists for a very long time in dead plants and purified virus preparations can remain viable in the laboratory for 5 years. The virus is not killed until heated for 10 minutes at 194°F or until treated for 24 hours with 60 per cent alcohol, but it is sensitive to soap. Probably tomato sap dried on clothes can remain infectious for a long time. Virus liberated from a plant into the soil can remain infectious for a long time if the soil is moist; freezing and drying the soil renders it non-infective. Infection from other plants and products. Common tomato mosaic is caused by the same virus as common tobacco mosaic, which is common in all tobacco-growing countries. Fermented, commercialj smoking or chewing tobacco may retain the virus in an infectious condition.

Hence chewing tobacco is a common source of infection in commercial tobacco production. Besides species of Nicotiana this virus can infect most plants belonging to the family Solanaceae and also species in other families, notably asclepias, snapdragon, larkspur, zinnia, beans, spinach and possibly celery. Potato, black nightshade and woody nightshade (bittersweet) show no symptoms, but they are carriers of infection. From these hosts the virus can be returned to tomato.

The nightshades afford a means of carrying it over winter as also do susceptible greenhouse ornamentals like Petunia, Solanum and others. There seems little danger to the tomato crop from mosaic-infected potatoes though there is evidence of one such a transmission out of doors.

It may be supposed that the virus multiplies more luxuriantly in vigourously growing plants.

The statement that a crop can be "manured out of mosaic" only means that suitable manuring may overcome the consequences of infection; the virus in the plants is not destroyed. When nitrogen is very deficient the mottle becomes invisible; if nitrogen is then supplied it reappears. The influences of light and temperature have been mentioned above. Attacks are often found to start near doors and ventilators, which can scarcely be due to draughts but rather to infection from outside the house.

SPOTTED WILT

Then a large number of tiny bronze-coloured spots appear

either in rings or scattered over the whole leaf surface. Growth stops for a time so that infected plants appear stunted. Young seedlings may be killed but more often the plant resumes growth after a time, but its leaves become distorted and have a rather conspicuous yellowish mosaic mottling. The fruit may be unaffected or it may bear pale-red, yellow or almost white markings, often in concentric rings. The virus is sap-transmissible but not to the same degree as tomato mosaic. It is not spread by aphides but only by thrips.

Curiously enough it is only in the larval stage that thrips can take up the virus from an infected plant, but once they have done so they remain infective throughout their life. Lycopersicum Virus 3 is not very resistant, it remains viable only a few hours in expressed tomato sap, is inactivated by drying, by oxidizing agents and by exposure for 10 minutes to a temperature of 108°F.

It has a very wide host range and commonly infects many popular greenhouse perennials which then form reservoirs of infection for the next tomato crop. Most of these show little sign of infection or bear only a few rather inconspicuous ring marks on their leaves.

TOMATO YELLOW LEAF

It starts as a well-marked mosaic mottle on the youngest leaves. The older leaves turn yellow, except along the main veins which remain dark-green. As the plant grows the yellow spreads until only the topmost leaves remain green. The whole picture resembles nitrogen starvation. Some evidence of seed transmission has been obtained. In properties the virus closely resembles that of common tomato mosaic. The aster yellows virus can also infect tomato and causes a certain amount of yellowing and a bushy type of growth, as it stimulates development of leafy buds and shoots in the axils of the leaflets as well.

AUCUBA MOSAIC

This disease is caused by a strain of the common tomato mosaic virus and must not be confused with aucuba mosaic of potato, caused by quite a different virus. Bright yellow spots often appear on the leaves of tomato plants infected with common mosaic. If these are punched out and used for inoculation the aucuba mosaic disease is produced. It is also sometimes found by itself in commercial tomato houses. About 5 days after inoculation the youngest leaves curl down and their surface becomes rough, wrinkled or corrugated.

By the seventh or eighth day yellow spots appear and become gradually larger, until in extreme cases the whole leaf surface is yellow, with a few raised islands of dark-green standing out on it like blisters. More often the yellow and green areas are of about equal extent and intimately intermingled. Plants are not killed by the virus but their growth is checked and the fruit may be mottled. Properties of the virus are the same as for common tomato mosaic. There is some evidence that seed-borne infection can take place in a very small number of seedlings, perhaps by virus dried on the seed coat.

TOMATO DISTORTING OR ENATION VIRUS

This is another strain of the common tomato mosaic differing in that, in addition to causing mottling of the leaves, it stimulates the plant to excessive and unnatural growth. The leaflets may be increased so that a fern-like effect is produced with very many small leaflets crowded together, gradually diminishing in size and ending in a corkscrew-like tendril. In some cases small outgrowths (enations) are formed on the under-side of the midribs. There is a tendency to excessive growth and production of giant plants crowded with foliage.

BUSHY STUNT

This is a distinct and apparently rather uncommon virus which is easily transmitted by sap but has no known insect vector. In seedlings it causes yellow and purple colouration of the leaves, yellow spotting and death of the young leaves and often a sunken brown lesion at soil level, causing the seedlings to fall over. In older plants it causes cessation of growth by the terminal shoot and stimulation of the side shoots, hence the name " bushy stunt."

The lower leaves turn yellow or purple and often there are concentric rings of bright yellow or purple, or purple lines along the veins of a yellow leaf. Ultimately the lower leaves shrivel and drop off. This virus is particularly interesting because it has been purified and obtained in the form of minute cubical crystals.

MISSHAPEN FRUIT

The tendency to production of ridged and misshapen fruit is an inherited one, as shown by Pape. It is also, however, influenced by conditions of cultivation. Thus, Kondine Red is a variety rather prone to bear ridged fruit but it only does so when the plants are set very far apart and richly manured. To avoid this disfigurement seed should be saved only from carefully selected plants with well-shaped fruit.

The seed crop should be grown under conditions conducive to misshapen fruit so that any inherited tendency to do so becomes evident. In normal commercial fruit production spacing and manuring should be that found by experience to suit the variety grown.

BLOSSOM-END ROT

The most striking symptom is a dark, more or less sunken, area at the blossom or style end of the fruit, i.e. farthest from the stalk. The trouble begins as a water-soaked spot at or near the base of the style, or the whole of that end of the fruit may look as though bruised.

The affected part soon becomes leathery and usually turns black. As the rest of the fruit continues to swell the blossom end becomes slightly sunken or remains abnormally flattened. Various bacteria and fungi are commonly to be found in the affected tissues and have from time to time been described as the cause of blossom-end rot.

They are, however, purely secondary invaders, living in tissue already damaged by a physiological disturbance. Especially if the surface of the fruit remains moist these secondary organisms may enter and set up an extensive rot. Tomato varieties in which the style remains attached for a long time seem particularly susceptible to blossom-end rot, perhaps because fungi and bacteria become established on the dead style and are hence ready to invade even a small spot of true physiological blossom-end rot.

The most important preventive measure is regular watering to avoid alternate drying out and soaking of the soil. It may be worth while to pick off the dead corolla and style on varieties that retain them long after the fruit has set. Otherwise fungi or bacteria established in the dead tissue may set up secondary rots following blossom-end rot. The variety Comet seems particularly susceptible to this trouble.

HOLLOW FRUIT

Fruits divided by a single septum into two compartments are particularly prone to develop empty spaces between the septum and the fruit wall. Such fruit when shaken often makes a rustling or rattling sound.

A larger or smaller part of the seed and flesh inside may be black and dead. This unfortunate phenomenon, which appears especially towards the end of the summer, is not due to attack by any parasite. It is due especially to a period of drought during the swelling of the fruit but other growing conditions may possibly cause it. In such circumstances when there is a heavy crop only part of it will contain normal contents.

GREENBACK

Greenback appears as a hard green or yellow area at the top of the fruit, surrounding the stalk. When ripening starts this part does not turn red like the rest but remains green and then slowly goes yellow or orange. It never becomes red and soft like the rest of the flesh. Greenback, like blotchy ripening, is due to a deficiency of potassium, especially in fruit exposed to strong sunlight. Tomatoes forced to ripen at a temperature exceeding 90°F do not turn red but yellow.

Lycopin, the red colouring matter, is not formed at such high temperatures. Greenback always occurs on the uppermost part of the fruit as it hangs on the truss, which is, of course, the part most exposed to the sun. When defoliating tomatoes one should be careful to leave enough leaves to shade the trusses during the hottest part of the day.

If the foliage is scanty, side shoots should be allowed to grow out and produce a couple of leaves before being stopped. Glasshouses should be shaded during the hottest periods. Care should be also be taken to supply enough potassium and water. Kondine Red and Tuckwood are very prone to greenback, Selandia much less so.

BLOTCHY RIPENING

In blotchy ripening larger or smaller parts of the fruit surface do not change colour with the rest as the tomato ripens. When it begins to turn red they remain green for a time but gradually turn yellow, develop a waxy appearance and a turnip-like flavour.

The vascular bundles in the blotches are often dark and show faintly through the skin as dark lines. The blotches may occur on any part of the fruit, a feature which distinguishes them from greenback which occurs only round the stalk. Blotchy ripening is seen mainly in June and July. Tomato varieties differ greatly in susceptibility to it; Kondine Red is one of the most susceptible.

Blotchy ripening is due to cultural factors or to diseased roots. In manurial experiments at Cheshunt it has been demonstrated that the greatest number of fruits with blotchy ripening occurred on unmanured plots and on plots deficient in potassium and nitrogen.

If tomato roots are diseased, if they receive too much water or too little, they cannot absorb enough nutrients and this may lead to blotchy ripening. The dampness of the subsoil should therefore be taken into account. It may well happen that the top 6-9 inches of soil are fairly moist but that the deeper levels are too dry. Tomatoes seem to develop most blotchy ripening in heavily limed soil.

In Danish nurseries blotchy ripening is often prevalent where there is a large amount of phosphoric acid in the soil. If a heavy dressing of manure is given the roots may be damaged and thus less able to absorb nutrients so that the plants may starve on over-manured soil. Blotchy ripening is usually worst in new houses where growth is unusually vigourous. In summers with plenty of sunshine less potassium and more nitrogen should be supplied than in dull, cloudy seasons.

CRACKED FRUIT

Cracking of tomato fruit occurs in two chief forms, single large cracks and many small ones; both types occur mostly on the top half of the fruit. The large cracks happen especially when a dry period is suddenly followed by a damp one. They are therefore usually found in the outdoor crop, where the grower has least control over the moisture relations of the plant. Cracking may also result from a sudden acceleration of growth for other reasons, as after manuring.

The skin makes poorest growth under cool conditions and then has difficulty in keeping pace with rapid expansion of the flesh. The numerous small cracks on tomato fruit occur especially in greenhouses where there are great fluctuations in atmospheric humidity. Growers should therefore be careful not to stop night heating too early It is probably the increased root development that counteracts the effect of a fluctuating water supply. Potassium permanganate is usually used for watering, etc.

ROLLING OF THE LEAVES

Tomato leaflets sometimes roll up so much that the upper surface becomes completely overarched by the upcurled margins which may meet above the midrib of the leaflet. Such leaves feel very hard and stiff. Varieties and individual plants may differ somewhat in the degree of leaf rolling. Tomatoes, perhaps because they are more extensively stripped than the glasshouse crop. The rolling of tomato leaves has no connection with the leaf roll disease of potatoes, which is due to virus infection. Rolling up of the top leaves may occur on young, actively growing plants. It occurs mainly when the crop has been heavily manured so that there is a high concentration of nutrient elements.

Such plants have dark-green foliage and the top may have a violet tint somewhat reminiscent of spotted wilt.

It occurs mainly in spring, particularly on tomatoes but also on other greenhouse plants such as pelargoniums and early forced potatoes. The outer cells become gorged with water, swell and burst the skin.

The swellings then have a white powdery appearance but gradually turn brownish as the broken tissue falls away. Affected leaves have a tendency to roll. Edema occurs when the roots are absorbing more water than the leaves are able to transpire. A warm, moist soil increases the rate of water absorption but damp air, poor ventilation and shade check transpiration from the leaves.

The condition may also become serious when, after a cool period with slow growth, the temperature rises and supply of water and nutrients to the foliage is suddenly increased. Such plants have relatively few leaves unable to cope with the large increased water supply resulting from the altered conditions. To check oedema the soil should be kept fairly dry and there should be thorough ventilation. This condition arises in plants that have became much more prevalent after the adoption received too much water or excessive nitrogen of petroleum oil sprays against red spider.

The or have experienced a sudden change in con-film of oil checks transpiration from the leaves, ditions when planted out. The grower often finds a few plants which times been noticed some time after severely early attract notice by remaining short because pot-bound plants have been planted out. Once the roots have grown out into the new than in normal plants. These should be rogued soil they are able to absorb more water than out before the crop is planted out as such can be dealt with by the rather stunted tops, plants are partially or completely unfruitful.

Some time after planting out the stems growth. As a rule there are only a few of these may become soft and flaccid; in bad cases the in each batch of seedlings. Like the short-plants fall over without any actual rot taking jointed plants they are practically sterile, place.

When such a stem is cut open the pith These varieties seem to be due to inherited is found to be wholly or partly wanting. Plants defects. They occur scattered amongst normal that have fallen over in this way may remain plants and should be weeded out as early as green for a long time but make no growth possible.

POTATO

POTATO BLIGHT

The attack on potato leaves starts as a rule at the edge or tip of a leaflet as a light-green spot, which quickly becomes brown and limp, and later blackish-brown and dry. Under damp conditions the brown spots have a marginal zone of fine white mould, composed of conidiophores and conidia of the fungus, which burst through the skin of the leaf 2-3 days after infection.

There are at first a few scattered spots, usually on the lower leaves; but others quickly follow and the whole potato haulm may be destroyed in about a week.

The diseased leaves have a characteristic sickly sweet smell which can often be detected at some distance from a diseased crop. Brown streaks develop too on the stems and even the flowers and berries may be attacked.

Potato blight on under-side of the leaf. Tubers attacked by potato blight show leaden or dull purplish-grey, slightly sunken spots or irregular blotches in the skin. When they are cut through, rusty-brown streaks are seen growing from the skin into the flesh. In time the whole tissue becomes destroyed by a rather dry brown rot.

Commonly, however, P. infestans is quickly followed by secondary fungi, especially species of Fusarium and bacteria, which obscure the symptoms of blight and may cause an evil-smelling soft wet rot to supervene. In all its stages the blight fungus needs oxygen and it cannot penetrate rapidly into the flesh by itself. The secondary invaders feed on the starch grains that the blight fungus has only superficially attacked. P. infestans conidia may be formed on diseased tubers.

The most important source of infection every summer is certainly slightly infected seed tubers. These may give rise to:

- Quite a normal sprout,
- Infected sprouts which soon die or
- Infected sprouts which come above ground, bear conidia and thus infect the lower leaves of the crop.

The potato blight fungus can also grow out from diseased tubers and small dead shoots into the soil where it lives for a time, and may produce conidia which are splashed on to the leaves by rain.

The appearance of the fungus in a crop from diseased seed tubers has been doubted but was seen by de Bary. Severely attacked tubers are readily destroyed by bacteria and millipedes, and sprout feebly or not at all. It is only those tubers which are slightly infected which are able, when planted, to produce shoots that come above ground. In June and July spread of infection from these diseased sprouts to lower leaves of the crop proceeds unnoticed.

These centres of infection form starting points for the incredibly rapid blazing up of the disease which occurs when conditions become favourable to it in July and August. It may be supposed that it will break out at many places at the same time if the conidia are carried a short distance by the wind. It may be of importance whether the early potatoes lie on the windward or leeward side of the late varieties. Sometimes one may notice that a hedge has afforded a certain amount of shelter and that blight has started opposite a gateway or other gap in the hedge. Small potato patches in town gardens become attacked at the same time as fields in the country.

In experiments with cultures in sterilized soil or manure, where the blight fungus encounters no competition from other organisms, it has survived the winter out of doors. The possibility of it overwintering in ordinary field or garden soil is very small but it must be remembered that there are always some groundkeeper tubers not killed by frost and the fungus may naturally survive the winter in some of those.

Heaps of chats discarded when the crop was graded or left on the site of an old pit are another possible source of infection. If the potato tops are killed early by blight and the tubers are left to ripen in the soil a very small crop is obtained but the tubers do not become infected, provided they were covered by a good thick layer of soil.

If blight attack comes late, retarded by weather conditions or by spraying, so that the tops are still green and bear fresh blight spots when the crop is lifted, there is great danger of the tubers becoming infected, especially in wet weather. Liability of the tubers to become infected depends partly on the variety, partly on the soil. Haulm and soil remain infectious for 10, possibly 20, days after the haulm has died. Haulm cut down by frost is at once freed from infection.

ALTERNARIA BLIGHT

This leaf spot disease is known and feared under the name "early blight" because it tends to appear earlier than true potato blight, called there " late blight." It is due to the mould Alternaria solani, formerly called Macrosporium solani Ell. & Mart., which causes leaf spotting also on tomato and egg plant. More or less circular brown or blackish spots appear on the potato leaves from July onwards. They are limited by the leaf veins though in severe outbreaks several spots may become united. Under a lens the spots are seen to be marked with darker, irregularly concentric lines, hence the name "target spot" sometimes applied to them.

In damp weather a fine olive-brown mould may appear on the surface. Some varieties show considerable resistance to the disease and only very small brown spots develop on their leaves. This is the form of Alternaria blight usually seen in Britain, where, indeed, the disease is regarded as of no economic importance and where conidia are seldom to be found on the spots.

It seems clear, in fact, that not all the "target spot" on potato leaves in this country is attributable to Alternaria infection. Very early spraying with Bordeaux mixture prevents an outbreak of Alternaria blight; according to one investigator Burgundy mixture is even more effective.

Where the disease has been prevalent care should be taken

that the tubers do not become infected by spores from the haulm at lifting, and to store them in a dry cool place. The optimum temperature for Alternaria rot of the tubers is 60°F and its growth ceases at 41°F. Alpha, Bravo, King Edward, Majestic, Kerr's Pink, Edzell Blue and Up-to-Date are all susceptible; Juli is fairly resistant and Golden Wonder more so. This fungus causes a minor leaf spot disease which in some parts of Europe causes premature withering of potato haulm in severe seasons.

The leaf spots appear from July onwards; they are at first yellow, then brown on the upper surface, and are tinged greyish-purple with the conidia on the under-side of the leaf. The routine sprays against potato blight control this disease as well.

POTATO WART DISEASE

Wart disease increases rapidly on land where potatoes are frequently grown, as in gardens md allotments, and in a space of 2-3 years it may attain such proportions as to ruin the crop. Direct losses of crop from the disease on an agricultural scale have been small but stringent quarantine regulations against it hamper export of produce from infected districts and discourage the cultivation of susceptible varieties. Hence the regulations which aim at excluding wart disease from countries in which it does not yet occur cannot be held to be unreasonable.

On several occasions the disease has been brought in with small consignments of potatoes among immigrants' luggage or by small-scale commercial traffic across a frontier. Only plants of the nightshade family, Solanaceae can be attacked by this fungus. Tomatoes develop swellings on roots and stem, and under experimental conditions small warts on the leaves also. Henbane and a number of species of Solanum develop small swellings on the roots and eventually small galls on the stems.

It has not proved possible to infect thorn apple (Datura stramonium], deadly nightshade (Atropa belladonna) or any kind of tobacco. In a wet season warts may be found on the stems and lowest leaves of potatoes, and even on the inflorescence, but as a rule the attack is confined to the tubers and is first noticed at digging.

The stolons may be affected but the roots are never attacked. As a rule growth of the haulm is normal and unchecked. The warts on the tubers start as tiny white granular swellings growing out from the eyes.

They quickly enlarge into cauliflower-like structures which may be larger than the tuber itself. Above ground they are light green, below ground at first white, then brown. Old warts decay and turn black and even at the normal lifting time many will be partially rotten, or break off and decay in the soil, leaving only a small scar on the tuber.

Occasionally new small outgrowths may develop in the clamp around the scars, giving a disease picture like that of powdery scab. Powdery scab infection of the sprouts occasionally produces swellings simulating wart disease, as may also irregular second-growth, covered with scab. Sometimes microscopic examination is necessary to distinguish between these diseases.

The swellings of crown gall also resemble those of wart disease but the former is known on potato only from artificial inoculations. The wart disease swellings contain a vast number of minute, brown, thick- walled resting sporangia which become liberated into the soil as the tissues decay. When they germinate a number of extremely minute, mobile swarm spores emerge and swim about until they meet a young potato tuber. They then penetrate the skin of an eye and settle down inside the potato cells. Here in a fortnight clusters of from 1 to 7 summer sporangia are formed which liberate hundreds more swarm spores.

These are partly responsible for multiplying the number of infected cells on and around the developing wart, partly for infecting other tubers. Second generations of summer sporangia form during the summer, and their development requires 10-20 days. They do not overwinter but are replaced in the mature wart by the thick-walled winter sporangia, formed after a kind of sexual process. These not only survive the winter but remain alive in the soil for very many years; they are not destroyed even when an infected potato is eaten and digested by an animal.

The formation of warts is due to the potato cells being stimulated to abnormal growth and multiplication by the fungus cells which invade them. The wart disease organism is one of the most primitive of fungi and never develops any kind of mycelium.

METHODS OF SPREAD AND REQUIREMENTS FOR INFECTION

The most important and most dangerous means by which infection spreads is the transport of seed potatoes, not only diseased tubers but healthy potatoes from infected land, which may carry resting sporangia in the soil adhering to them. Potato peelings in kitchen waste added to compost heaps or manure and dung of animals fed on raw potatoes may also carry infection.

The fungus may be carried from infected land in soil on carts, tools, boots, and the feet of domestic animals and poultry. Soil particles transported by storms, or floods, or by crows and other wild birds may also spread the disease.

Resting sporangia can germinate over a wide temperature range (40 to 86°F) but best at 66-68°F. Presence of oxygen and nitrates encourages germination. Invasion of the host cells by the swarm spores also occurs over a wide temperature range (45-86°F), best at 60-68°F, and oxygen is needed for this process. Soil reaction seems to have little effect on infection. It occurs over the whole range between pH 3.9 and 7.5, even pH 5 is said to be quite favourable.

Soils more acid than these are seldom cultivated and one can feel no assurance that infection will not occur on soils more alkaline than 7.5. No essential difference has been found between infection on clay and sandy soils. Extreme drought may completely prevent outbreak of potato wart disease.

On the other hand the warts become particularly large with a heavy fall of rain. Plants that mature early or form no tubers owing to virus disease, etc., may escape attack, as may very early varieties. Resting sporangia in the soil survive longest when the ground is under permanent grass or is left unbroken for other reasons.

After a severe outbreak infection has been proved to persist in the soil for 12-16 years. Wart disease is easily and completely controlled by growing immune varieties. No potatoes from an infected crop may be sold for planting and no tubers visibly affected with wart disease may be sold for any purpose. Only approved immune varieties of potato may be grown on land where wart disease has occurred at any time.

In a scheduled Protected Area the only potatoes allowed to be planted are tubers from a crop certified to conform to official standards of purity and health and to have been grown on land free from wart disease.

It is also forbidden to remove rubbish or compost from such places or to dispose of manure. Export of tubers from boroughs where infected areas occur is also prohibited. There is no practicable method of destroying wart disease resting sporangia in a field soil. On a very small scale soil can be disinfected by steaming; baking the soil is not effective. Treatment of soil with 5 per cent formalin for 24 hours also kills the sporangia.

Incorporation of sulphur in the soil also has some effect but only in large quantities which make the soil sour and are unreasonably expensive. Implements used on infected land may be disinfected by dipping in 2 per cent formalin solution, but disinfection of potato " seed " has not proved effective. Cooking of potatoes used as fodder helps to prevent contamination of manure and is, in any case, required by the Wart Disease Order as far as infected tubers are concerned.

COMMON SCAB

The disease called common scab of potato is caused by the ray fungus Actinomyces scabies, which also attacks beet, carrot, radish, swede and turnip. Care is needed to distinguish between it and powdery scab. The ray fungi are generally regarded as being somewhat intermediate in structure between bacteria and fungi. Several hundred species occur in soils, where they play a large part in the breakdown of decaying organic matter; others live in water and a few cause disease in animals. Potato scab appears as rough, scabby spots on the otherwise smooth skin of the tubers. The spots may grow together and the whole surface of the tuber may be covered by them.

If careful examination is made when the layer of cork begins to form in the skin the first signs of disease will be detected as small grey-brown spots under the lenticels where the fungus is stimulating the living cells to production of additional layers of cork. The outer cork is burst open and the new cork follows suit, producing the scabby outer surface of the lesion. The more prone the variety is to cork formation the scabbier it will be.

Roots and stolons of potato may be slightly attacked. Tubers that protrude above ground are not affected, nor do scabs increase in the clamp. Tubers are the most readily infected; if they have a long growing period, the period during which they become scabby is increased, hence, perhaps, the various reports that " spraying induces scab."

A strong earthy smell due to the presence of the actinomycete is peculiar to common scab; powdery scabs have no such smell. The fungus may overwinter either in the soil or in the scabs on tubers. It can survive on cellulose and similar plant products in the soil for years at a time. Several instances have been reported in which disinfected tubers planted in freshly broken forest soil produced scabby crops.

On the other hand an interval of several years between potato crops has sometimes led to a striking reduction in the amount of scab. The fungus reproduces simply by its fine hyphae breaking up to form spores, easily disseminated in soil and water. Manure may be regarded as infected if it comes from farms where scabby potatoes are used raw as fodder. Infection may also be spread by compost and, of course, by planting infected tubers.

VARIETAL SUSCEPTIBILITY

No potato variety completely resistant to common scab

is known and on soil favourable to the fungus all varieties are attacked to about the same extent. Early varieties tend to escape infection, probably because the cork layer in their skin is formed before the soil becomes warm enough to suit the fungus. Main crop varieties do show some differences in susceptibility, probably associated in part with the size and texture of their lenticels.

Breeding for scab resistance is in progress in Minnesota where the possibility of physiological races of the fungus is also under investigation. Disinfection of the tubers is effective in destroying the scab fungus on them but is only of value when they are to be planted on scab-free soil, or in soil treated with sulphur or otherwise disinfected.

Digging in grass cuttings may be effective in gardens or allotments provided they are available in sufficient quantity. Up to a stone per square yard has been recommended. On a field scale ploughing in rye or other green manure in the spring may be helpful. Consistent use of sulphate of ammonia instead of chili saltpetre will gradually reduce the amount of scab, but one must be careful that the soil is not made too acid for other crops that need lime.

Combined use of dung and sulphate of ammonia has proved helpful. The proprietary preparation Bayer P. at the rate of about 1 oz. per square yard or 54 Ib. per acre, dug well in before the potatoes are planted, may greatly reduce both common scab and stem canker.

It works very well when used with sulphate of ammonia. Liming may be necessary because of other crops in the rotation, but it is well to have the soil analysed and to apply only the lime requirement. Where it is important to obtain a crop of smooth tubers a piece of ground may be kept unlimed for potato production. Continuous cropping with potatoes is liable, however, to lead to trouble from other diseases and especially from root eelworm.

POWDERY SCAB

Powdery scab, sometimes known also as corky scab, first appears on the skin of potato tubers as pale little pimples.

Either before or after lifting, the skin over the pimple splits and falls off, exposing a snuff-brown powder composed of the spore balls of the fungus.

For a considerable time the individual powdery scab remains as a shallow crater-like hollow, filled with brown powder and fringed by the loose edges of burst skin. Under good cultural conditions the scab seldom exceeds, but in potatoes grown in wet land or which have made considerable second growth, large open scabs and cankers may be formed, especially on the ends and protuberances of the tuber.

In such extreme forms of the disease microscopic examination may be needed for confident diagnosis. Corticium solani may be found in the wounds and various other secondary organisms may enter and set up a rot. In exceptional circumstances the fungus may continue its activity on tubers in the clamp and stimulate them to production of irregular tumour-like outgrowths closely resembling those of wart disease. Spongospora subterranea also attacks the roots of potatoes, and less often of tomato, causing irregular white nodules.

It is doubtful if these cause any damage to the plant but they contain spores which are liberated into the soil when the roots decay and thus add to the reservoir of infection in the ground. S. subterranea is one of the most primitive fungi, a relative of the fungus which causes club root in cabbages. Its spores, which are stuck together in peculiar sponge-like balls, can remain alive in the soil for 4-5 years at least and are easily spread to new localities in soil on tubers or by adhering to roots of transplants. When they germinate they liberate swarm spores which penetrate either the roots through root hairs or the tubers through lenticels or small wounds. Cultural conditions.

Powdery scab is found on both sides of the Atlantic and in the potato-growing districts, but is only of importance where there is a heavy rainfall and on ill-drained soils. In dry soils a reasonably clean crop may be lifted even though grown from infected tubers.

No varieties are immune from the disease but there is

some difference in susceptibility, said to be correlated with the thickness of the cork layer in the skin of the tuber. Liming is said to make the trouble worse, and application of sulphur at the rate of 3-6 cwt. per acre checks it.

Where the disease is prevalent, soil drainage should be made as efficient as possible and the rotation should be arranged to leave 4-5 years between potato or tomato crops. Soil disinfection is impracticable in the field. Diseased tubers should not be returned to the land but may be cooked and used for fodder.

Manure from animals fed on raw infected potatoes will contain undamaged spores and should not be used for potato fields, nor should potato peelings and similar kitchen refuse be put in the compost heap. Disinfection may be practised where infected tubers have to be used on land free from the disease. They may be dipped for 3 hours in formalin solution, 1 pint of formalin in 30 gallons water, or 1: 2,000 corrosive sublimate solution. Moistening the tubers and rolling them in powdered sulphur may also be effective. Spores of closed scabs still in the pimple stage may not be killed by any of these methods.

Secondary Scab Injury

Both in common and powdery scab, especially when associated with second growth, the damage is often increased by the activity of a number of insects and other pests, such as the larvae of Sciara flies, wood lice, and millipedes which penetrate the outer layers of the tuber flesh by way of the scabs and cracks and convert it into a dry crumbling crust. The stem eelworm Anguillulina dipsaci produces extreme crust-like rotting of this kind. Secondary fungi may also play a part; even dry rot often starts at a powdery scab lesion. When storage conditions are too damp, or too warm, soft-rotting bacteria obtain easy access to the damaged tissue and the whole tuber decays.

BLACK SCURF AND STEM CANKER

This, one of the commonest and most cosmopolitan of

fungi, was first recognized in connection with potatoes, with which it is almost invariably found. But it also attacks many other plants, cabbage, tomato, wheat, lettuce, asters, though sometimes in the form of a specialized strain more or less adapted to a particular host, and has been found on many weeds such as buttercup, shepherd's purse, field thistle, plantain, etc. It can also live freely in the soil.

More recent studies have made it possible to sub-divide this very large genus into a number of smaller more clearly denned genera.

When this is done C. solani falls in the genus Pellicu-laria and some authors think it identical with a fungus described earlier under the name Hypochnus filamentosa Pat., hence it has been re-christened Pellicularia filamentosa. This name has not yet, however, been generally accepted.

On potato tubers one often finds hard scurfy patches to which the soil adheres. When the tubers are washed these patches show up as dark-brown or black spots and lumps. They are the sclerotia of C. solaniy sometimes described as "The dirt that won't wash off." They are attached to the skin by very fine hyphae and can be scratched off with the finger nail, leaving the uninjured skin of the tuber beneath.

Apart from making ware potatoes a little unsightly this phase of the fungus's life history does no immediate damage. Occasionally it has been found to attack tubers through the lenticels before they are lifted, causing small shallow brown pits rather like those of pit rot. When "seed" tubers carrying C. solani sclerotia are planted, however, the fungus becomes active again and many cause more or less serious injury to the sprouts. It may be found on the young sprouts even in the chitting boxes but most damage is done after planting. Then sunken brown spots appear on the sprouts below soil level. Or the tip of the shoot may be killed before it emerges above ground; it then puts out a side shoot which may escape or may also be killed.

In the worst cases groups of such dead shoots are found with the result that blanks occur in the potato drill. Later in the summer the fungus causes dry, brown, sunken wounds on the underground parts of the stems. If these are deep and girdle the stems the haulm may be weakened and the leaves roll, somewhat as in primary leaf roll though less stiffly.

When the starch cannot reach the tubers, green aerial tubers are produced in the axils of the leaves. Brown lesions also appear on the stolons, which may be killed. The plant then forms merely a cluster of small angular tubers round the base of the haulm. The roots too may be attacked, especially the fine absorbing rootlets.

In all these brown sunken lesions only the mycelium of the fungus, or occasionally Rhizoctonia sclerotia, can be found. But at the base of vigourous haulms where the air remains moist the fungus covers the stems just above soil level with a white powdery film, sometimes extending up them to a height of 6 inches. This film consists of a layer of basidia and basidiospores. No harm seems to be done to the plants by this sporing stage of the fungus but it offers special possibility for its spread by means of wind-blown spores. The effect of an attack on yield of tubers varies greatly from practically nothing to severe losses where the fungus has caused extensive gaps in the drills.

Conditions of Growth

Potato sprouts may be infected partly from the see"d tubers, partly from the soil. The fungus seems to thrive in all cultivated soils, is not much affected by soil reaction and can grow over a wide range of temperatures. The rate at which the tubers sprout determines the severity of the disease in spring because the damage is greater when potatoes are planted early in cold soil; nevertheless early planting generally pays! As a rule lavish use of dung and artificial manures helps the crop.

VERTICILLIUM WILT

The symptoms of Verticillium wilt in potato usually develop only after the haulm has attained a normal size, though in extreme cases it may be stunted, with all the leaves rolled, and yields a negligible crop. More often, from July onwards, the leaves become limp and turn yellow at the tips and edges. Often one side of the leaf turns yellow and dies. Finally the leaves die and hang down brown and shrivelled against the stem, so that the whole plant gradually dries from below upwards.

Characteristic of the disease is the brown stain in the woody strands, often extending throughout the whole haulm. In autumn a delicate velvety white mould of conidiophores may appear on the base of the dying plants, and finally the stem may be blackened by the minute resting bodies of the fungus. Tubers on infected plants are usually also infected from the stolons, and if one slices them at the heel end the vascular bundles will be found to be brown, often far into the flesh. Brown vascular bundles occur also in Fusarium wilt, black-leg and other diseases, including leaf roll, and some forms of slight frost damage.

There is no natural separating layer between the tuber and the stolon so that when the crop is lifted a small wound is inevitably made at the heel end which takes a little time to heal. Before the cause of potato leaf roll was understood it was frequently confused with Verticillium wilt and there was consequently a tendency to exaggerate the prevalence and importance of the latter disease.

EFFECT OF ENVIRONMENT

Verticillium wilt is widespread on both sides of the Atlantic but in warm climates it tends to be superseded by Fusarium wilt. High temperatures check the growth of the fungus but the drought and excessive transpiration accompanying them usually kill all the infected potato plants. Experiments have indicated that heavy nitrogenous manuring increases the susceptibility of the crop; soil reaction has little effect. Provided a normal rotation of crops is practised the disease is unlikely to become prevalent. Crops containing many infected plants should be used for ware only.

If a few plants show Verticillium wilt in a crop grown for " seed " they should be dug out and burnt, preferably with those on each side in the drill as otherwise a few infected tubers may be left and lifted with the crop. Also neighbouring plants may have become infected by root contact and if this occurs late in the season they may not show distinctive symptoms before the haulm begins to ripen off.

FUSARIUM WILT

It is associated with a brown discolouration of the vascular bundles throughout the plant, as in Verticillium wilt; indeed, the two diseases can only be distinguished with certainty by isolation of the fungus in pure culture. The FusaHum enters the tubers along the vascular bundles from the stolons and in the worst cases it grows through them all round the tuber so that a brown ring is visible when it is cut in half. During the winter the fungus slowly grows into the flesh and may cause rotting of the whole tuber.

If those only slightly infected are planted they give rise to infected shoots which grow only into small stunted plants. Fusarium oxysporum requires a fairly high temperature; it makes scarcely any growth at 48°F and its optimum lies between 77° and 86°F. In cool climates it is only likely to be troublesome in potatoes forced under glass. Warm, damp soils favour infection but drought hastens the death of the infected plant. Trouble is not likely to arise even in forced potatoes if they are raised from " seed " from a normal field crop. If the disease does occur potatoes should not be grown in the same soil for several years.

BLACK DOT

This very common fungus may damage potato haulm in dry seasons. The stems are attacked just below soil level, and become marked or even girdled by dark-brown or blackish lesions from which the skin breaks away. Girdling of the stems leads to yellowing and rolling of the leaves and premature dying of the haulm.

The fungus is common on the tubers where it causes silvery grey patches covered with tiny black dots like those of silver scurf. It may also girdle the stolons and penetrate some distance up the stems along the vascular bundles.

The tiny black sclerotia, sometimes bearing conidia on their surface, may be found on any of the diseased tissues. King Edward, Red King and Doon Star are considered to be susceptible varieties. The fungus is cosmopolitan and has been described under several other names, of which Vermicularia variant Ducom. is that most used on the Continent. It is regarded as a serious parasite in France but is of little consequence in cooler climates. The fungus is better known as a parasite of tomato and egg plant.

DRY ROT

During storage of potatoes numerous species of Fusarium may gain access to the flesh of damaged tubers, sometimes through wounds but mainly through tissues already killed by blight. Dead patches of tissue caused by sun scald or damaged by disinfectant or other chemicals may also afford entry to weak parasites. Susceptibility of the tubers to these rots increases during the winter. True dry rot, however, is caused mainly by one easily recognized species of Fusarium coeruleum, so-called because its conidial pustules instead of being salmon pink have more or less of a blue tint.

They may be covered with a white or pinkish spore powder but if this is rubbed off the blue colour will be found in the base of the pustules. It also sometimes appears in the mycelium, lining cavities in the rotting flesh. F. coeruleum is an active parasite though it may often gain access to the tuber through a scab wound or a bruise.

The first sign of dry rot is a brown patch on the skin, which rapidly increases in size, becomes slightly sunken and surrounded by concentric wrinkles. The underlying flesh is deeply penetrated by a soft brown rot, full of cavities lined with white or blue mycelium, and giving off a peculiar sweetish smell. In time the rot destroys the whole tuber which becomes hard, dry, shrivelled and very light in weight. The characteristic conidial pustules burst through the dead skin.

SILVER SCURF

Sometimes silvery patches of skin are noticeable on potato

tubers, especially when they are washed. The underlying tissue may be quite normal or a little darker than usual. Silver scurf is seldom seen at lifting time but develops in the clamp and may be regarded as a slight blemish rather than a disease. In wet clamps the condition may be widespread and extensively affected tubers may tend to shrivel if stored afterwards in a dry place.

Otherwise no real harm is done. Under warm, damp conditions a black velvety layer of short conidiophores and conidia appears on the silvery patches, springing from very tiny black specks in the skin, rather like the tiny sclerotia of Colletotrichum atramen-tarium. No control measures are normally called for, but if the fungus becomes undesirably prevalent a longer interval between potato crops should be tried, as infection of the tubers is derived from the soil.

PINK ROT

The fungus, which occurs in the soil, may attack the underground parts of the haulm and the stolons, sometimes causing the plants to wilt, but this usually happens only at the end of the season. Losses from the disease are due to infection of the tubers..

This occurs normally at the heel end while they are still attached to the parent plant, but the fungus may also sometimes gain access through an eye or a wound. Inside the tuber it causes rather a rapid rot but the diseased potato remains fairly firm until it is completely killed. When squeezed, however, sap generally oozes from it.

If an infected tuber is cut and exposed to the air the flesh turns pink in a few minutes and within half an hour the whole diseased portion becomes salmon-coloured. Later it slowly turns purplish-brown or blackish-brown. Similar colouration may occasionally occur in healthy tubers or in those attacked by black-leg.

The infected tubers decay fairly quickly and are not likely to survive until planting time, hence there is little risk of spreading the disease with the " seed." As the causal organism is a soil-borne water-mould, thorough drainage of potato fields is desirable and a long rotation should help to prevent infection of the soil being built up to a dangerous level. All infected tubers should be destroyed, not dug or ploughed back into the land.

JELLY-END ROT

This is a physiological disorder associated with a form of second growth, especially in Arran Pilot. When second growth occurs evenly over the rose end of the tuber, so that it becomes abnormally elongated, the tissue at the heel end loses its starch and becomes soft, flabby and shrunken. Ultimately it usually shrivels up and falls off leaving an irregular scar.

WET ROT

Even at lifting, tubers may be found that are completely soft and rotten; as a rule they are from a black-leg plant or have been suffocated by flooding or puddling of the soil. Potatoes that have been attacked by blight and black-leg or have been damaged during lifting and bruised by rough handling in transit, packed tightly in railway vans or in a ship's hold, or stored under ill-ventilated conditions, may be transformed in a few days into a rotten mess. Numerous species of bacteria may play a part in this decay.

If heating in the clamp is discovered in time and ventilation is supplied many of the tubers will merely develop black-heart. If not, and the temperature continues to rise, the potatoes will be suffocated and afterwards reduced to a soft, stinking, slimy mass by the common putrefactive bacteria that cannot attack living tissues.

VIRUS DISEASES

Virus diseases occur wherever potatoes are grown, though they only spread rapidly in climates favourable to their insect vectors, hence potatoes grown in some mountainous countries and in exposed wet maritime districts can be kept free from viruses without great difficulty. In commercial potato fields the virus-infected plants occur scattered throughout the crop because they arose from tubers which were already infected when they were planted. Secondary spread of the diseases within the crop is usually evident only late in the season or only in plants grown from the harvested tubers the following year. In lowland districts the decline in productivity resulting from virus infection would render potato production uneconomic if fresh " seed " tubers were not secured every few years from a suitable " seed " producing region.

Tubers from a diseased plant mostly contain the virus or viruses that have attacked it but a few may sometimes escape infection. Once a tuber has become infected, it, and the plant grown from it, always remain diseased. Hence the primary source of potato virus disease is the planting of diseased tubers. It is impossible to tell from its appearance or by any chemical test whether a tuber is virus-infected or not.

The symptoms caused by infection by a potato virus vary greatly, partly according to the variety of potato infected, partly according to the time elapsed since infection (symptoms in the second years are often very different and more severe than those in the season of infection) and partly according to environmental conditions, for heavy manuring, especially with nitrogen, and warmth may " mask " the symptoms in the haulm.

Carriers, varieties which can be or always are infected with a particular virus without showing any symptoms, are common among potato varieties. In practice potato virus diseases are usually referred to one of four categories, irrespective of the particular virus present.

These are leaf roll, mild mosaic, severe mosaic and aucuba mosaic. This is the most destructive of all potato virus diseases and the most difficult to control. Rolling of the leaves may also be caused by drought, waterlogging, fume damage, Corticium solani, black-leg, insect damage to the stem, eelworm and the like, but in all these cases the rolled leaves are limp or at least pliable, and usually it is the upper young leaves that are affected.

In true leaf roll, derived from a diseased tuber, the lower leaves are rolled, point more or less upwards and are stiff and hard. Mild mosaic is defined as a mild light and dark green mottle, in which the leaves are neither distorted nor reduced in size. It may be so faint as to be visible only on a cloudy day or when the leaf is held to the light. Often it is clearest in the young haulm, and less apparent in late summer and autumn. Mottling may also result from manganese or magnesium deficiency, but then the whole haulm is usually affected throughout a field or in considerable parts of it. Aucuba mosaic, so called from a fancied resemblance to the variegated leaf of Aucuba japonica, is recognized by clearly defined, bright yellow spots scattered over the lowest leaves.

The leaves are not distorted and are of normal size, and the plant seems to suffer little from the disease. Somewhat similar variegation may result from spring frosts. This includes all diseases of the mosaic type in which the surface of the leaves is roughened, crinkled, streaked with dark lines, distorted or reduced in size and in which the haulm is more or less stunted, especially in the year after infection.

OTHER VIRUS DISEASES OF THE HAULM

Paracrinkle is a peculiarly severe form of crinkle caused in the stock when a scion of King Edward potato is grafted on certain other varieties. All King Edward plants contain the para-crinkle virus without suffering any ill effect from it. Paracrinkle is purely a laboratory disease; it has never been seen in the field and no means of transmitting it other than grafting has yet been discovered.

Top necrosis begins by the appearance of numerous small brown dead spots on the youngest leaves. These rapidly spread and the shoot dies back from the tip. If any tubers were formed before the plant died, corky spots develop in them during storage and they either die before planting or, at best, give a weak shoot which soon dies.

Top necrosis is very seldom seen in a field crop but it is of practical importance in two ways. In the first place different potato varieties react by top necrosis to different viruses. Hence grafting scions from a plant under examination to a selected number of indicator varieties is a standard method of identifying the viruses it contains. This method is particularly valuable in detecting "carried" viruses.

Secondly, because the progeny of a plant that reacts by top necrosis are all killed, this is a self-eliminating disease. A virus cannot be perpetuated in a potato variety in which it causes top necrosis. Hence much attention has been paid to breeding new potato varieties which react by top necrosis to as many viruses as possible.

VIRUS DISEASE OF THE TUBERS

Net necrosis is the name given to a fine brown network in the flesh of the tuber. This is a reaction of certain varieties, notably Golden Wonder, in the year they become infected with leaf roll. That is, tubers with net necrosis produce plants with primary leaf roll but these do not, as a rule, yield more tubers showing net necrosis. A somewhat similar brown network may be the result of slight frosting of the tuber.

Tuber blotch refers to small brown spots or blotches in the flesh of the tuber, rather like those known as internal rust spot. The condition may be a result of infection with Solanum Viruses. Potato varieties all react in much the same way to Solanum Virus, the cause of leaf roll, though the symptoms are more pronounced in some, e.g. President, than in others like Great Scot.

Their reaction to Solanum Virus 2 also differs in intensity rather than in kind. They react very differently, some by top necrosis, others by a mild mosaic or by showing no symptoms of disease at all.

RED PEPPER

LEAF SPOT

The spots are at first watersoaked but soon dry out and are then white with a dark-brown margin. When many spots occur on one leaf it turns yellow and drops. Spots may sometimes occur also on the leaf-stalks and fruit stalks. The fungus cannot penetrate the skin of the fruit but may grow into it through the stalk and then sets up a " stalk rot." It is likely to appear sooner or later where the plant is only grown on a small scale as a glasshouse crop.

Spots appear on the fruit, starting as small translucent or dark-red depressions, which quickly grow larger and gradually become lighter in colour. The pustules of the fungus form a dark coating on the diseased tissue and exude pinkish masses of conidia, which are spread by rain splashes or on clothes or tools of workers.

Sap Warts

When pepper is kept wet, especially if the atmosphere is nearly saturated or when great changes in humidity occur, it easily develops oedema. This takes the form of a large number of very small grey-brown warts on the underside of the leaf. One should try to avoid excessive humidity or much change in the moisture supply to the plant.

Grey Mould

We are not sure whether Botrytis is the primary cause of the damage or net. Quite likely the fruit suffered first from blossom-end rot and the grey mould followed and increased the damage. Leaves and stalks may also be attacked. The plants must not be kept too dry or the fruit will develop blossom-end rot, or too damp or Botrytis will be encouraged. Indeed the art of growing this, like other greenhouse crops, consists in supplying the plants with enough water in a damp enough atmosphere without letting them become moister than is necessary.

BLOSSOM-END ROT

Capsicum fruits may develop blossom-end rot like tomatoes. The first sign of the disease is a more or less sunken area at the tip, or close to it, on the half-grown fruit. The spot is usually a little paler than the surrounding healthy tissue but it may be dark-green and translucent.

Gradually the spot grows larger but it does not cover the whole of the blossom half of the fruit as may happen in tomato. The spots may, however, become infected by fungi or bacteria, which set up a rot that may involve the whole fruit. The spots are not primarily due to a parasite, however, but to a physiological disturbance. Higgins has studied the problem closely in Georgia, both with controlled watering and under natural conditions out of doors.

In all cases it was shown that drought caused the appearance of the spots. In the open air they become infected by bacteria or fungi in the course of a few days. In fairly young spots neither fungi nor bacteria were found.

Only as the spots increased in size did these organisms invade them and thrive on the already damaged tissue of the fruit. The spots start by collapse of the large, thin-walled cells round the ends of the vascular bundles. Red-pepper plants should not be exposed to drought, especially when the fruit is half-grown and has been growing vigourously.

BROAD BEAN

CHOCOLATE SPOT

Chocolate spot is one of the commonest diseases of broad beans and field beans. Dark brown spots appear on the leaves and stems. They may be small, more or less circular, and scattered, in which case little harm is done, or they may spread rapidly, cover the whole leaf surface and cause decay of the stems and death of the plants.

In severe outbreaks very few pods mature. Brown spots occur on the pods as well and may penetrate to the ripening seed. Typical chocolate spot may be caused by the ubiquitous mould Botrytis cinerea, but there is also a specialized species, B. fabae, peculiar to broad beans, which causes a similar disease. Chocolate spot is favoured by moisture, ill-drained soil, shade and crowded planting. Hence it is a much more disastrous disease of field beans than of isolated rows of beans in gardens. It is also encouraged by potassium deficiency or by malnutrition of any kind.

MINOR LEAF SPOTS

Cercospora fabae Fautrey causes small chocolate-coloured spots or so across on broad-bean leaves. Under favourable conditions they become larger with a sunken grey centre and a raised dark-brown margin. In wet weather both sides of the spot may bear a silver-grey layer of conidia. Similar spots may be caused by the pycnidial fungus Ascochyta fabae Speg. This fungus may also cause light sunken spots with a broad dark-brown margin on the pods and is known to be seed-borne. Both Cercospora fabae and Ascochyta fabae are widespread but neither is usually considered important enough to warrant control measures.

SCLEROTINIA DISEASE

Broad beans, and especially field beans, are susceptible to a stem rot similar to that caused in other plants by Sclerotinia sclerotiorum. In the case of these beans, however, the cause of the trouble is usually a distinct fungus, 5. trifoliorum var. fabae, very closely related to the fungus of the well-known red clover rot. Although the race commonly found on broad beans has been distinguished from S. trifoliorum by small technical differences it can also cause rotting of red clover, and the typical S. trifoliorum can cause rotting of broad beans. Both fungi also attack common vetch, garden pea, sainfoin, lettuce, and possibly other plants.

BEANS—DWARF AND RUNNER

Anthracnose

Except in the warmest and driest parts of the earth dwarf and runner beans are liable to surfer severely from attack by the anthracnose fungus, the conidial stage of which is only too well known.

It has become fairly common especially in cool wet summers, but the ascospore stage has only been found. All kinds of dwarf and runner beans may be attacked. The disease is seed-borne and the seedlings may show brown stripes on their stems. In damp weather new spots readily arise, extending lengthwise into sunken brown stripes.

On the leaves the spots generally start on the under-side of the veins, though the surrounding leaf tissue may also be killed. The most conspicuous symptoms are those on the pods, which are circular or elliptical sunken brown spots with a darker edge. From these the fungus grows into the seed, in which it remains alive until germination, and then attacks the seedling. On dark-coloured seeds it is difficult to see the spots resulting from infection. On white-seeded varieties the larger spots are pale brown but small ones may escape notice. The fungus overwinters in the soil, especially on dead pods and other plant fragments. Plants attacked in the seedling stage or soon after may be killed outright; at later stages the damage is mainly to the pods.

The fungus fructifies as minute pinkish pustules of slimy spores on the surface of the spots and these are scattered, especially by splashes of rain and by handling when the pods are being picked. Enclosed localities where the air is apt to be moist and still favour the disease.

Manuring with potash and superphosphate may make the plants a little less sensitive; sulphate of ammonia should only be used on a table crop, not in seed production. Bean varieties differ greatly in susceptibility so that considerable success has been attained in developing resistant types. Resistance may be dependent on one or more inherited factors but so far these have always been dominant in hybrids. It soon became evident, however, that the fungus, too, exists in numerous races different in infective capacity.

Hence a bean variety may be resistant to several races of the fungus but not to others. Hence new varieties have to be exposed to infection by all known races of the anthracnose fungus before they can be described as resistant, and preferably this property should be combined with resistance to other diseases like halo blight, rust and mosaic. So successfully has this been done that some bean varieties now contain three hereditary factors for resistance and are resistant to 34 races of the anthracnose fungus. Actually they all become infected but in resistant varieties the fungus only penetrates the skin and then dies.

Fusarium Wilt

The chief symptom is a rolling inwards of the leaf edges,

followed by wilting of part or the whole of the plant. The fungus occurs only in the vascular strands which are stained dark brown as in Vertidllium wilt, especially at the base of the plant. It is apparently not seed-borne and infection presumably comes from the soil, through wounds made by soil insects or by the mechanical splitting of the stem base by secondary roots. The optimum temperature for its growth is 79-82°F.

Foot Rot

Foot rot is caused by a soil-borne fungus prevalent in the Evesham district of Worcestershire. It attacks the roots and stem bases of dwarf beans, turning them reddish brown. The leaves turn yellow and only the first-formed pods swell properly. Foot rot causes most loss in dry seasons, when it may reduce the yield by one half. Strong-growing varieties like Flageolet are somewhat resistant as they put out fresh roots from the lower part of the stem to replace the diseased ones. The same fungus has been found to cause a similar disease in French beans grown under glass. In this case sowing in sterilized soil or peat is recommended, followed by protecting the stem base with the rim of a bottomless pot when the seedlings are planted out in the house.

Rust

This is found on dwarf and runner beans all over the world. The cluster cup stage of this rust is white. This is usually followed by cinnamon-coloured pustules of uredospores on both sides of the leaf and finally by blackish-brown pustules of teleutospores. Bean rust is only occasionally reported. Sprays and dusts give no control of rust and if it breaks out the diseased plants should be burnt or composted and replaced by a resistant variety.

Red Nose

When beans are germinated in the seed-testing laboratory a slight pink discolouration is sometimes noticed at the micropyle or germ pore. This has been called " red nose " and is due to infection with the common fungus Pleospora herbarum. This fungus cannot cause disease of the growing crop but the infected seeds often decay or give sickly seedlings that die early. If they survive this stage they grow into normal healthy plants.

Treatment with one of the proprietary organomercury seed dressings often gives improved germination. Beans harvested under unfavourable conditions, or otherwise of poor quality, may be attacked during germination by bacteria and moulds like Penicillium, which reduce the percentage germination.

Treatment with an organo-mercury seed dressing or with red copper oxide may give a definite improvement in germination. Some beans germinate and develop their cotyledons (seed leaves) normally but the terminal bud or plumule is seen to be lacking. An axillary bud of one of the cotyledons may grow out to replace it but only gives a poor plant. The cause may be bacterial infection or damage at threshing, which has cracked the embryo just above the attachment of the cotyledons. Microscopic examination will decide which cause is involved.

Finally, beans may germinate badly (and cook badly) after a very hot, dry autumn or after storing in a dry warm place which hardens the skin of the seed. Storage in a damp warm place also brings about certain enzyme reactions within the bean and leads to formation of a " hard kernel," which also reduces germination and cooking quality.

Halo Blight

A number of bacterial diseases are known to affect dwarf and runner beans but their symptoms are usually somewhat similar so that they can only be distinguished by bacteriological investigations. The disease is usually attributed to Pseudomonas phaseolicola. This is seed-borne, and infected seed gives rise to stunted, spotted or yellow deformed plants which usually soon die. From them the bacteria are spread, probably largely by rain splashes, to healthy adjacent plants. On the leaves small, irregular, watersoaked spots appear, that are at first pale, but later turn brown with a transparent halo-like margin.

Similar spots may appear on stems and petioles and from them small drops of bacterial slime may ooze out. When the pods appear they too become spotted and sometimes deformed, or even killed. Seeds from infected pods are likely also to be infected with the bacteria. Sometimes they are spotted, wrinkled, or shrivelled, but they may bear no obvious mark of infection and yet give rise to an infected seedling.

It too is seed-borne and infected seedlings may be killed by it as they emerge above ground. On older plants it forms minute watersoaked spots or pale-green wilted areas which enlarge, dry out and turn brown. They usually have a yellow border. Reddish streaks appear on the stems, or reddish lesions may girdle the stem at the level of the cotyledons, or one of the lower nodes. The stem easily breaks at these points. On the pods there are dark-green, watersoaked spots that become dry, brick red and sunken.

LETTUCE

RING SPOT

This fungus causes oblong sunken spots, at first watersoaked, then yellowish and finally reddish or light brown, like small patches of slug damage, on the stem and the under-side of the midrib of lettuce leaves. Similar spots appear on the leaf blade, where they may be round or angular, defined by the larger veins.

In later stages of the disease the tissue in the centre of these spots dies and falls out, so that the leaves look as though perforated by hail. The attack starts on the outer leaves and spreads inwards. Endive is also attacked by this fungus, so is the wild lettuce Lactuca scariola and probably the weed Crepis. Bleaching powder is well mixed with ten times its weight of water and left to settle for half an hour.

The clear liquid is then strained through muslin and the seed steeped in it for 4-6 hours, dried and sown at once. A small quantity of a wetting compound may be added to the solution. Other control measures include rotation of crops and destruction of the remains of all infected crops after the marketable heads have been cut. In greenhouses care should be taken to keep watering at a minimum and to avoid splashing soil on the plants. Seedlings may be sprayed with a copper fungicide, but this is impracticable with maturing lettuce as the copper deposit on the leaves would render it unfit for human consumption.

OTHER LEAF SPOTS

The first symptoms are small light-brown spots on the leaves, usually round but sometimes angular or oblong. The spots enlarge, join up and may cause large parts of the leaf to become brown and wrinkled. On the outer leaves the centre of the spots may fall out so that they look like hail-damaged plants. Often the spots have a dark-brown edge or bear a number of darker concentric lines. Septoria lactucae Pass, causes light-yellow irregular blotches, which turn brown and become covered with minute dot-like pycnidia.

In later stages of the disease the spots may join together, or fall out as described for Pleospora above. The fungus is seed-borne and its pycnidia can be detected on infected seed by examination under a binocular microscope.

DOWNY MILDEW

Lettuce out of doors in frames and in greenhouses is often attacked by downy mildew. The upper surface of diseased leaves is marked by rather vague pale or brownish spots, the under-side of which is covered by a thin whitish mould. When conditions are favourable the fungus may spread very rapidly. It has a low optimum tempe-rature, 43-50°F, and is hence favoured by cool weather and by low temperatures in houses and frames.

Young plants are more susceptible than older ones. The fungus may overwinter in the soil but seems unable to survive severe frost. Morphologically similar downy mildews occur on many weeds of the lettuce family (Compontae) but it is unlikely that most of these can attack lettuce. Two races of downy mildew on lettuce, one of which infects the wild lettuce Lactuca virosa while the other does not. Both occur on L. serriola. It is often impossible to establish where the infection of cultivated lettuce comes from, but one must remember that the conidia can be spread from a diseased crop in many ways, through the air, by adhering to boxes and baskets used for lettuce and on clothes or boots of workers. Seed-borne infection has not been demonstrated.

GREY MOULD AND RED LEG

Grey mould may cover a greater or smaller portion of a lettuce head and cause it to rot. This is especially likely to occur if the head has been damaged in some way and if there is a little dead tissue about, as from marginal spot or tip bum. In forced lettuce and outdoor white lettuce much more serious losses may arise by infection of the stem of young plants. This causes a reddish sunken wound, " Red-leg" which rapidly kills the plant, especially in overwintered seedlings transplanted in spring. Later in the season the stem wounds may heal and the plant mature more or less normally. Dusting with Folosan is said to afford some protection.

SCLEROTINIA DISEASE

These fungi gain access to withered leaves at the base of the plant and spread rapidly from them into the living head, which after a couple of days becomes limp and appears as though boiled. The heads may also appear to have fallen over, or to have been cut at the base. These fungi are easily recognized by their profuse development of white mycelium and their large, black sclerotia.

The former species will attack almost any herbaceous crop, the latter has already been found on beans, asparagus, radish, etc. The clover rot fungus, common in red clover fields, also occasionally attacks lettuce. The control measures described for S. sclerotiorum are applicable to all these species.

RHIZOCTONIA DISEASE

Symptoms. When the heads are formed small rusty or

chocolate-coloured spots appear on the under-side of the midrib of those leaves in contact with the soil. The midrib and leaf blade soon wither but the stem of the plant is less susceptible. Wilting of the outer leaves on sunny days is one of the first signs of the disease. Gradually the heads decay completely; at first they are brown and slimy, then they dry up and turn almost black. The brown mycelium of the fungus with its small dark irregular crust-like sclerotia permeates the dead tissues. The disease is favoured by damp and even more by high temperatures.

MOSAIC

The first symptom of lettuce mosaic is a clearing of the leaf veins, making them appear more prominent. Later a rather vague irregular mottling with shades of green may develop or the whole leaf may become yellow and hard-looking, as though damaged by wind or frost. The leaf surface may be abnormally crumpled and its shape modified. There may also be tiny brown spots between the veins, or brown lines along the veins, and scorching of the leaf edges. The whole head is dwarfed and does not heart properly.

When the crop " bolts " plants infected with mosaic remain stunted and may have brown spots on the stem and flower head. In cos lettuce browning of the veins is rare, but scorching of the leaf edges often occurs. The virus is transmitted by lettuce seed and is spread in the field and glasshouses by aphides, including Myzus persicae and Macrosiphum gei. It can infect groundsel and sow thistle (Sonchus asper) and also sweet pea and culinary pea, causing a " streak " disease in the former and mottling of the leaves in the latter. It appears unable to infect dandelion.

TIP BURN

Shrivelling of lettuce leaf edges may be due to infection by Pseudomonas marginalis (Brown) Stapp and scorching to potassium or phosphoric acid deficiency. Probably more common than either is the physiological disorder known as tip burn. The disease begins as small brown spots near the edges of the leaves. As a rule they occur where the finer veins come together from the leaf edge. These small spots seem to hinder water from reaching that portion of the leaf which lies beyond them, hence it withers, turns yellow and dies so that a brown band appears along the edges of the leaves.

Factors conducive to the production of dead leaf edges are: excessive moisture, great variations in the amount of moisture, lack of light, high temperatures and overmanured soil, especially an excess of nitrogen. Vigourously growing plants are more likely to develop the condition than those growing more slowly. According to experiments, lettuce is sensitive to an excess of salt in the soil. Good results have been obtained by using fertilizers free from chlorine, viz. monoammonium phosphate and potassium nitrate.

Experiments have also shown that varieties rich in sugars are less susceptible than those with a low sugar content. Steam sterilization of the soil predisposes to tip burn and that it is worst in heated houses and after prolonged dull weather. These factors are all conducive to soft growth, and tip burn sets in when the plants are exposed to strong sunlight

BEET

RUST

Beet rust is one of those which has all its stages on the same host plant. The orange-coloured cluster cups (aecidia) appear in May on the under-surface of leaves of young plants, and occasionally on the leaf stalks. This stage is seldom seen on seedlings but more often on the young leaves produced by overwintered roots kept for seed production. Spores from the cluster cups spread to young beetroot plants, on which successive generations of uredospores appear as rust-coloured pustules all through the summer.

Towards the close of summer they become replaced by pustules of the dark-brown teleutospores. Beet rust is most prevalent on mangolds and sugar-beet, but occurs also on garden beet, spinach beet and the wild sea beet (Beta maritimd). The most important control measure is to keep seed crops as far away as possible from fresh sowings. It may also be as well to inspect seed crops at regular and frequent intervals in spring as long as the cluster cups appear and remove the leaves that bear them.

They should at once be put in a pail of lime water or other disinfectant fluid to prevent spores from being spread about the crop. Mother roots for seed crops should, if possible, be chosen from plants free from rust and stripped of leaf stalks which mighl bear the fungus.

LEAF SPOTS

Ramularia beticola Fautr. & Lamb, and Cercospora beticola Sacc. cause very similar circular leaf spots with a whitish centre and dark-purple margin on beetroot. On spinach beet the spots are pale with brown edges. When necessary, spraying or dusting with a copper fungicide should be practised. A leaf spot of mangolds caused by Stemphylium botryosum Wallr.

BLACK-LEG

The young seedlings turn brown at the soil level and shrivel up. Black-leg is worst on wet and sour soils which should be improved. Various fungi may play a part in the disease, including species of Pythium in the soil and Phoma betae Frank which is seed-borne. Hence treatment of the seed with the standard seed disinfectants may have a beneficial effect.

BEET MOSAIC

Mosaic is very general in seed crops of mangold and sugar-beet, but seems less prevalent in garden beet, where it is only found in individual plants with conspicuously mottled leaves. In our infection experiments red beet proved not very susceptible and spinach beet even less so.

Hence attacks are only to be feared when these crops are grown in the vicinity of sugar-beet or mangold seed crops. Beet mosaic will, however, attack spinach and hence overwintered spinach may be a source of infection from which the virus may be spread to beet by aphides. In our experiments the common goosefoot was not infected but wild sea beet did develop mosaic.

HEART ROT, DRY ROT AND CANKER

The innermost leaves turn black and die and gradually all the leaves on the plant may wither. Later new clusters of healthy leaves may appear on the crown outside the dead heart. On the root a slow, dry brown decay starts at the shoulder and gradually penetrates the tissues. In garden beet black spots may appear in the flesh.

Heart rot and dry rot are much better known in sugar-beet and mangolds than in red beet. In the former they are known to be associated with limed soils and with dry years.

Of late years excellent control has been obtained by manuring with borax at the rate of about 20 Ib. per acre. On soils where previous experience has shown that beet is liable to suffer from trouble of this kind borax should be applied in spring. Care must be taken not to exceed the recommended rate or the plants will be poisoned. Not more than 1 oz. to 12 square yards should be given. Watering with very dilute borax or boric acid solutions may check an outbreak that is just beginning.

BEET YELLOWS

Yellows is a common disease in sugar-beet and mangold crops where it affects all the plants in large patches or over whole fields. The virus may be spread by aphides to garden red beet, spinach beet and spinach and to goosefoot. Infected spinach beet develops stiff yellow leaves. Leaves of infected red beet are also stiff, but the yellow colouration tends to be masked by the red pigment in the leaves.

Leaves of the diseased plants are gorged with starch, and are often prematurely attacked by the black mould (Alternaria) which normally only appears on the dying leaves in autumn. The disease is not carried by the seed and there is no infection from the soil. The virus overwinters in beet plants, especially in seed crops, and is spread from them by aphides. As far as garden beet and spinach are concerned it seems to be prevalent only in the vicinity of sugar-beet crops.

CUCUMBER

ANTHRACNOSE

The seedlings may be attacked and fall over but this seldom happens. Leaf spots are the usual symptom at all stages of growth of the crop. They are pale green and transparent, later becoming dry, reddish brown, with a pale margin and tend to fall out. They enlarge more quickly than those of blotch and are more apt to fuse together into large withered areas, which may kill the whole leaf. Sunken dry patches, later becoming dry and floury, appear on leafstalks and stems. Sunken pale spots also appear on the fruit.

The diseased tissue commonly bears pinkish coni-dial pustules of the fungus, easily distinguishable under a magnifying glass. In greenhouses that have contained an infected crop the fungus occurs in the soil, on the glass, especially at joins, on decaying wood, old paper, etc.When the disease has occurred the houses should be thoroughly cleaned after the crop has been removed, sulphur should be burnt in them, all crop debris destroyed, the woodwork and glass should be sprayed with formalin 2:100 and the soil should either be renewed or thoroughly disinfected.

GUMMOSIS

When the seed has been sown in infected soil the seedlings may be attacked. As a rule the first symptom is a water-soaked spot on the hypocotyl below the seed leaves and often a dark stripe extending as far as the latter, on which dark spots may also develop. If the attack is severe the stem cracks at the attacked places. Under favourable conditions the infected seedlings may survive but the fungus is easily spread further from these lesions. In older plants oblong spots with yellowish-brown margins may develop on the stems and leaf stalks. On the leaves the fungus causes yellow or greyish-yellow spots which burst open.

They are larger and more irregular than the small regular spots characteristic of blotch. The disease on the fruit is the most general, most characteristic and best-known phase. Depressions occur in the skin, but several such spots often fuse to form large irregular patches. The spots often exude a gummy liquid, which hardens into a yellow or brown bead, entirely or partially filling the depression. This characteristic gummy discharge gives the disease its common name " gummosis".

The conidia of the fungus form a grey or greenish-black layer over the depressions and on the edge of the spots. The disease is caused by the mould fungus Clados-poriurn cucumerinum. Races of this fungus isolated from many different places varied greatly in their capacity for infection. It can over-winter in the soil, on woodwork, glass, etc. Cucumber seed extensively but was unable to demonstrate seedborne infection; its possibility is not, however, to be excluded. The fungus can grow at temperatures between 32° and 85°F but thrives best about 70°F, whereas cucumber plants grow best about 80°F

If the atmospheric humidity is kept below 85 per cent development of the fungus is greatly checked. The fruits may also be attacked. There the spots are at first small, dark and somewhat sunken, not marked off from the surrounding healthy tissue by a sharply defined border. They quickly increase in size and may sometimes cover half the upper side of the fruit or more. Often the diseased tissues sink in to form a constriction in the fruit. This usually happens at one end of the cucumber; if it is at the stalk end naturally the entire development of the fruit stops.

MILDEW

Mildew may be very serious on outdoor cucumbers, it is of general occurrence but usually less serious on the hothouse crop. The disease begins as small white spots which may gradually coalesce to form a continuous powdery white coating over the shoot tips and leaves, particularly on the upper surfaces. When mildew is found in a greenhouse the ventilation should be improved and the old leaves removed as soon as possible. The remedies for mildew may also be tried usual.

BACTERIAL LEAF SPOT

This disease first appears as round or irregular transparent spots on the shoots and later on as well-defined, watersoaked, brown leaf spots, which dry up and fall out, just as in blotch. In the morning one often sees on the under surface of the spots small, clear, bacteria-filled droplets which dry to leave a chalky deposit. Watersoaked spots yielding a slimy white discharge may also arise on stems and leaf stalks. On the fruit the spots are at first very small, round and watersoaked, but later turn chalky white at the centre as the tissues die and dry out.

The disease is spread at pruning, by water splashes, and perhaps by insects.

It can be transmitted on the seed, which may be disinfected by soaking for 5-10 minutes in corrosive sublimate solution, following by rinsing in clean water and drying. This disease is due to a bacterial infection of the vascular bundles and effects cucumbers, melons and gourds. The leaves develop pale green spots that quickly spread and are followed by a general withering of the leaves and shrivelling of the stems. When the latter are cut a sticky, whitish, bacterial slime exudes from the bundles. This is a serious disease of outdoor cucumbers where it is spread by the knife at cutting and by insects, especially by a leaf beetle.

CHLORINE POISONING

Cucumbers have often been injured by being given stable manure, or latrine contents, to which chloride of lime had been added as a disinfectant. The leaves suddenly develop yellow areas at the tips, along the edges, or between the main veins. The newly set fruits suddenly stop growing and are spoilt; often they shrivel from the tip.

Generally the injury appears at irregular intervals, i.e. it appears suddenly, then an improvement occurs, followed by another bad period, and so on. When one realises that the damage is due to injury to the roots by chlorine from the chloride of lime one can understand that the first outbreak follows application of the manure.

Then when the chloride has been partly washed out the plants recover until damage to the roots again takes place. One should make sure that chloride of lime is not added to the manure or latrine contents that are to be used. If one discovered that it has been added one should water the plants very heavily to try to wash it out as quickly as possible. A tendency to bitterness in cucumbers is unfortunately quite common, especially amongst those grown in hothouses, but it has also occurred in cucumbers grown out of doors for sweet pickles. The bitterest fruits are dark outside and yellow inside but in most cases one cannot see from the outside whether a cucumber is bitter or not.

The bitterness begins at the stem end and spreads through the flesh from there. There may be a noticeable difference in taste between two slices cut close to one another. A cucumber which is not bitter when pickled may become so later. Bitter cucumbers are most frequently met with from June to August. On the whole the smooth types are far more liable to bitterness than the spiny-skinned kinds.

This is well known to many housewives and it is, therefore, easier to sell a spiny variety than a smooth one.Various factors encourage bitterness in cucumbers. There is some difference of opinion as to which is most important but it appears that sudden changes in growing conditions are of great importance. The phenomenon is often encountered when a period of sunshine follows a long period of cloudy weather, when the plants are too dry, or when they are very severely pruned.

Other factors predisposing to bitterness are: damage to the roots by parasites, over-watering, and use of too strong manure or of fertilizers containing chlorine, wide variation in atmospheric or soil temperatures and, finally, setting of too many fruits. Individual plants may vary greatly in their tendency to bear bitter fruits.

One should endeavour to maintain steady growth by

supplying heat during cold nights, ventilating and shading in strong sunlight, watering and cutting at the appropriate times, and avoiding too strong manuring. Bitter cucumbers are stained deeply by potassium iodide solution which may be useful in detecting them.

SHRIVELLED CUCUMBERS

Frequently cucumbers on hothouse plants cease growing and shrivel when about 2-4 inches long, without any obvious cause. The phenomenon is most frequent at high temperatures, with a moist atmosphere and wet soil. It also seems to be worst on plants carrying many large fruit. Several other conditions may stop growth of the young fruits: damage to the roots by chlorine, too strong farmyard manure, too cold water, drought, draughts, big ranges in temperature, or heavy pruning, leaving too few leaves so that transpiration is reduced.

STOPPAGE IN ROOT GROWTH

The roots may be damaged, so that they turn brown and no new roots are produced for a time, if too fresh manure or manure containing chlorine is used, or if the plants are watered with very cold water. This check to root development may cause injury to the fruits or of the leaves. A root rot of cucumbers, associated with species of Fusarium, the primary cause is believed to be bad cultural conditions, such as lack of drainage and poor aeration of the soil.

BLIND SHOOTS

Cucumbers in houses sometimes develop "blind shoots", i.e. the shoots suddenly stop growing and the tip ultimately withers, when side shoots develop to replace it. Even if these succeed, there has been a check to growth and, moreover, some side shoots may be destroyed also, so that it is some time before the plant but usually these are not the cause of blind shoots. The trouble is not due to parasites but to unsatisfactory conditions of growth which cannot always be precisely identified. It may occur where there is a draught from a hole in the glass, or from the ventilator.

Good results have been obtained by putting cones over shoot tips exposed to draughts.

The most frequent cause, however, seems to be that the plants were growing too vigourously, i.e. they had been given too strong manure, especially an excess of nitrogen and also too much warmth. Blind shoot tips are therefore most frequent on plants with large dark-green leaves. The larger the plants are when planted out the better. Large plants have a shorter time in which " the top can go " than small ones that take longer to attain their full height. The plants should not receive extra nitrogen until they have begun to set fruit. Temperatures during growth should not exceed 86°F.

DROOPING LEAVES

Cucumber leaves may develop transparent withered edges and in the course of a few hours the condition may spread over the whole leaf, which collapses as though scalded. The stalk remains normal. In some cases the leaves may have been noted beforehand to be pale yellowish-green between the veins; in other cases there was nothing peculiar about them. This phenomenon is almost confined to the spring months. No parasite has been found associated with it, and after a time it ceases to occur, though affected plants suffer some check in growth. In one instance the damage seemed to be associated with the use of too fresh cow dung in the beds.

Circumstantial evidence indicates that the trouble is due to some cultural fault. Cucumbers are a forced crop, grown at a high temperature and richly supplied with both food and moisture. If, therefore, they are checked even for only a short period, by a sudden rise or fall in temperature, or if the roots are damaged by too warm or too cold water or by too fresh manure, they receive a shock which finds expression in some kind of malformation of growth.

SUDDEN WILTING

It frequently happens, generally in August, that outdoor cucumbers suddenly lose their turgidity and wilt. The trouble was particularly prevalent, when rain came after a long dry period. There was no sign of disease on the root stalk or leaves except that the plants drooped and eventually withered altogether. In a locality where the matter was examined more closely the phenomenon was equally prevalent in soil where cucumbers had been often grown and where there had been none for 10-15 years.

Attempts to correlate appearance of the disease with some cultural condition in the various districts were unsuccessful except that everywhere the plants were said to have withered " after rain". In some places they had been watered during dry periods and still many plants withered. Those that died were about to set fruit; they seem to be particularly susceptible to the trouble at that stage of growth.

Floral Abnormalities

It happened at various places in Jylland that the outdoor cucumbers bore only male flowers or only an abnormally small number of female flowers. At one place where only Danish pickling cucumbers are grown, those raised in the greenhouse and planted out gave at first both male and female flowers and set normal fruit, but after a definite period the internodes became shorter, the leaves darker and only male flowers were produced.

Some plants sown out of doors, which therefore developed later, produced only male flowers, while those sown in the hothouses and kept there remained normal throughout. In another district the plants were of different varieties but the ovaries were not normal except on one or two and these set only a few fruits.

Seed of the same varieties and consignments gave only normal plants in a third district. From a place where there were only a few female flowers, seeds were saved from the few fully developed fruits produced, and these were sown in the following year alongside seeds from normal plants. Seed from the abnormal plants gave almost exclusively abnormal plants while the others were normal. Thus the abnormality was inherited but its first appearance was due to some external condition. It may perhaps have been a sudden spell of sunshine after a dull period with low soil temperature that caused the appearance of male flowers.

MUSHROOM

BROWN BLOTCH OR BACTERIAL SPOT

Mushrooms grown in too warm and moist an atmosphere may be destroyed by purplish-brown spots on the cap, in which slimy drops may be found by closer examination. These drops teem with bacteria, which may be spread about the crop by various insects.

Mushrooms -may be destroyed by this disease before they attain the size of a pea. The houses must be well ventilated after watering and the temperature should not rise above 55°F. Diseased fruit bodies should be removed and the attacked parts of the bed watered with chloride of lime.

BROWN PLASTER MOULD

Occasionally large crusts of felted mycelium appear on the surface of the manure and of the casing soil and slowly turn brown, owing to the production of minute sterile resting bodies by the fungus. These look like tiny yellow or brown sand grains. Brown plaster mould is introduced with the manure. If the manure is good and in good condition mushrooms will be produced in infected beds and even in the brown mould patches. On the whole, however, presence of the mould seems to reduce productivity of the bed.

GREEN PLASTER MOULD

On examining unfruitful mushroom beds one sometimes finds that the white mycelium of the mushroom has become surrounded by that of another fungus, which is at first also white, but soon turns yellow and ultimately grey-green. The fungus comes in with the manure and is said to be able to spread in sour wet manure. It may also be found on the surface of the beds. Direct control measures seem hopeless.

WHITE PLASTER MOULD

This fungus has a world-wide distribution and is of great

importance in mushroom cultivation, especially where wet or too old manure is used. Then the fine white mycelium spreads in great quantity over the beds until their surface appears to have been powdered with chalk. Old patches of the fungus may take on a faint pinkish or greyish hue. Whether or not cropping of the mushrooms is affected by the plaster mould apparently depends on whether the latter succeeds in invading much of the bed before the mushroom spawn gets a start.

WHITE MOULD

This is a parasite of the fruit body in mushrooms and other gill fungi. Diseased fruit bodies are often transformed into unrecognizable lumps or, if the infection occurs comparatively late, they have a short thick stalk and hardly any cap. Their whole surface is covered by a close white mould and they soon break down into a stinking rotten mass. The term " bubbles " has also been used for this disease. The fungus is said to be introduced with the casing soil, and use of fresh field soil from arable land is recommended. Local outbreaks in a bed can be treated by watering with formalin.

The brown race of cultivated mushroom is resistant to white mould. The white mould layer of the fungus contains two-celled resting spores and sometimes also a conidial stage something like that of Verticiilium.

The surface of the diseased caps was dark brown and the underlying flesh was stained yellow. Dactylium dendroides Fr. is a fungus attacking mushrooms and wild toadstools, which covers them with a close downy mycelium often with a pinkish tinge, and causes them to rot. It, too, probably comes from the casing soil.

VERTICIILIUM AND CEPHALOSPORIUM DISEASES

Several species belonging to these allied genera of moulds have been described as causing disease in cultivated mushrooms. Sometimes it causes shapeless, deformed mushrooms, covered with a fine white short mould, sometimes greyish-white spots on mushrooms of a normal shape, sometimes only white patches on the gills or stalk. This difference in symptoms depends on the stage of development at which the mushroom becomes infected. Even the most deformed fruit bodies do not rot but become firm and somewhat leathery in texture.

Conidia of the Verticiilium are produced in vast numbers in minute slimy heads on the mould and are spread about the crop by Sciara flies and other insects. The fungus does not attack mushrooms at temperatures over 70°F and seems unable to infect the white-capped strain. It causes mushrooms to become deformed, spherical, or pear-shaped bodies like those resulting from mycogone attack, but dry, brown and elastic in texture.

Gill Mildew Disease

All the above described species of Verticiilium and Cephalosporium cause spots on the surface of mushrooms and, in favourable circumstances, reduce them to dry, misshapen lumps. The present fungus only attacks the gills, which tend to fuse together and have a mildewed appearance. Affected mushrooms are still edible but do not keep well, as their gills tend to rot. Stable manure inevitably contains the spores of many kinds of fungi.

Many of these are killed by the high temperature during composting, and if the mushroom spawn is introduced to the bed at the correct moment it usually grows away ahead of the remaining fungi.

But under unfavourable conditions and if there are other sources of contamination, as from boots or tools, infection of the bed by competing fungi may forestall growth of the mushroom mycelium. Competition of this kind has already been mentioned under the heading of brown and white plaster moulds. Between the mycelium of species of Fusarium, especially F. oxysporum Schlecht. and F. solani (Mart.) Sacc. var. martii (Appel and Wollenw.) Wollenw., and the mushroom mycelium there is, however, a reciprocal hostility or " antagonism." Growth of one mycelium excludes that of the other even at some distance.

Presence of these Fusaria may lead to individual

mushrooms becoming brown inside and tough, or to many small button mushrooms remaining undeveloped. Sometimes the mushroom growth may be restricted to a single crop and sometimes there is no crop at all. Sterilization of the casing soil by heat kills these moulds and hence prevents trouble from that source.

MUSHROOM BED INVADERS

The false truffle Pseudobahamia microspora Diehl and Lambert produces nodular, cream-coloured or pinkish hard fruit bodies, buried in the casing soil. Xylaria vapouraria Berk., called " seaweed " by the growers, forms tough, branched, black resting bodies in the casing soil. Its white mycelium permeates the underlying compost. The sclerotia send branches up to the surface of the casing soil, where they form pinkish bodies flush with the surface. The sclerotia are said to have a strong smell like cucumber, detectable by crawling over the surface of the beds and smelling the casing soil.

ROSE COMB (NON-PARASITIC)

After oil sprays have been used to destroy insect pests, or where oil stoves are in use, or woodwork has been coated with creosote, mushrooms sometimes bear a malformation on the cap somewhat like the rose-comb found in some breeds of poultry. The irregularly lobed excrescence is due to abnormal gill formation on the upper side of the cap as well as on the lower.

Prevention of Mushroom Diseases

Production of a healthy crop of mushrooms depends largely on the care taken in preparing and tending the beds.

The usual disinfectant is burning sulphur, but there are other methods whose relative effectiveness has not been fully ascertained. Empty houses may be sprayed throughout with formalin i: 50, preferably with a power sprayer, so that soil, walls and ceiling are thoroughly soaked. The house is then locked and kept warm for two days and nights, then thoroughly ventilated.

The man who sprays should wear rubber gloves and his eyes and mouth should be protected. Boards and other woodwork may be disinfected by spraying or washing with copper sulphate or one can go over it with a blow lamp so as to burn off all loose splinters without scorching the wood. The copper sulphate treatment has an advantage over the blow lamp and formalin in that it leaves a small deposit of copper in the wood which preserves it without harming the mushrooms.

Milk of lime is also sometimes used on woodwork. If the compost beds are laid directly on the ground the latter may be covered with a layer of peat dust which may be watered with acetic acid at the rate of 2 gallons per 40 square yards. This acidified peat will prevent infection of the bed from the ground. Shelves used for mushroom growing may be washed in formalin and flamed with a blow lamp before the compost is placed on them.

The quality of the manure is of the utmost importance even in disease prevention. The right consistency of the horse manure, not too fresh and not too old, fermented long enough but not too long, can only be learned by experience. The manure should be moist enough to hold together when squeezed into a ball but not so wet as to drip liquid when this is done. The moisture can be increased by watering, or reduced by mixing with ground gypsum. The final fermentation (heating) should take place inside the mushroom house, and when it is at its height sulphur should be burnt to destroy fungi and insects on the surface. It is important to keep down mites and insects because, in addition to doing damage themselves, they spread fungus spores.

The casing soil should be obtained from the subsoil. Surface soil contains weed seeds and often Mycogone and if it must be used it should be disinfected by formalin, or by steam sterilization. A sandy loam is the best type of soil and a neutral reaction is desirable. Steam sterilization sometimes reduces the yield of mushrooms, while after using formalin the house must be left to ventilate for some time. The spawn is added to the compost as soon as the temperature 3 inches below the surface has fallen to 80°F, so that the mushroom mycelium may spread ahead of competitors. Diseased mushrooms should be removed with their mycelial strands and with the surrounding soil as well.

Before removing them the spot should be watered with formalin i: 50, chloride of lime or raw carbolic acid. It is better to remove diseased fruit bodies wholesale than to see the whole crop destroyed by disease. The diseased material should be burnt. Gypsum dusted over the beds helps to prevent plaster mould. If watering becomes necessary it should be done through holes in the casing soil instead of by sprinkling the surface.

The house should be kept damp by watering round about the beds. It is better to water between picking periods rather than during them. If the soil has been very mouldy one may spray it with very dilute Bordeaux mixture. When greenhouses are cleared after a crop the earth or manure removed should not be put near mushroom houses or on the site where mushroom compost is to be made.

ONION

SMUT

By careful examination smut can be detected on quite young seedlings. It appears as opaque dark spots or streaks under the skin of the first and subsequent leaves and sometimes in the bulb scales of young onions. The leaves first attacked are often abnormally thickened and twisted. Sooner or later the skin over the streaks splits irregularly, exposing the black spore powder. These spores contaminate the soil and many remain alive there for twenty years or more. Many of the infected seedlings are killed outright by the disease; others survive and grow, but are malformed, while still others grow almost normally but develop fresh smut lesions on their leaves and bulbs.

Individual plants may completely recover from the attack; diseased leaves and bulb scales are thrown off, and the bulbs ripen more or less normally, though they are always small.

The smut fungus which causes this disease can grow and infect seedlings at temperatures between 50° and 85°F. High temperatures after sowing therefore favour the fungus. Infection takes place from the soil and can occur only in the seedling stage, from about the second day after germination until the first leaf is developed.

Older plants cannot be infected. The fungus enters that part of the seedling which is beneath the soil or just above its surface. It penetrates the epidermis and permeates the tissues as far as the growing point, after which the new leaves become infected as they develop. The incubation period is 5 days. Wherever possible onions should be sown in soil free from smut spores. When they are big enough to be transplanted they can safely be transferred to infected land. It can attack many species of Allium and no resistant varieties of onion or leek are known.

Various methods of dry seed treatment have been tried to avoid the troublesome formalin treatment. Good results were obtained with Brassicol but the method should be tested in larger-scale experiments before being generally adopted. The onion seed was slightly damped and then mixed with half its weight of Brassicol.

NECK ROT

It may spread in storage. The diseased scales become somewhat brown and transparent, looking as though cooked, and in a moist atmosphere they develop a short hairy coating of grey mould. Bacteria often follow as secondary invaders and set up a rapid soft rot. In the later stages of decay small black sclerotia form on the diseased part or even over the whole bulb.

They may sometimes form a continuous scab, especially at the neck of the bulb. The disease is usually controlled when the bulbs are dried thoroughly and quickly after lifting and are then stored in a cool, dry, well-ventilated place. If they remain damp for long, neckrot may be very destructive. It is advisable to bend over the tops before growth has ceased. Onions are best stored at 35°F. Onions that have made very luxuriant growth seem more susceptible than those grown at a normal rate. White-scaled varieties are more susceptible than those with coloured scales.

DOWNY MILDEW

This disease does not usually break out until the bulb is about to form. Then a pale coating of the fungus appears on leaves and stalks. As the leaves wither the fungus turns brownish purple. In damp weather this coating may appear before there is any discolouration of the infected leaves, but they soon become pale and translucent and shrivel from the tip downwards.

In mild attacks one may mistake this yellowing for normal ripening. Gradually the infected leaves wither to the base, often being replaced by new ones which are also usually attacked. The dead parts are often covered later by a nearly black layer of the mould Stemphylium botryosum. In severe attacks the bulbs remain small, shrivelled or unripe. The disease also attacks shallots and in them it appears first on those grown from infected bulbs.

The leaves are small and wither quickly, becoming pale, flaccid and covered by the greyish-purple layer of mildew. From these plants the disease spreads to the rest of the crop and causes symptoms like those described on onion. Both on onion and shallot the fungus grows down from the leaves into the bulb scales and persists there until the following year. Thus shallots and onion sets are a common source of infection. In severely infected bulbs the outer scales become brown, translucent and shrivelled. Some may die during the winter; others survive and sprout prematurely.

Within the tissue of the infected leaves resting spores may be found, which can survive the winter and even remain alive for 5-6 years in the soil where onion tops have decayed.

Cause and Importance

Onion downy mildew is caused by the fungus Peronospora destructor, formerly often known as P. schleideni Unger. It can attack all varieties of onion and shallot, but is seldom seen on leek. When onion sets are used it is important to obtain them from a completely healthy crop. If that cannot be done the sets should be graded and only the large sound ones planted and these should be carefully polished.

The fungus in the sets may be killed by keeping them for more than 8 hours, or up to 24 hours, at a temperature of 104 degree F.

In some experiments good results have been obtained by dipping sets in copper sulphate solution, or lime water, but there were no untreated controls for comparison. There is much to indicate that the fungus penetrates the bulb scales too deeply to be completely killed by such treatment. Seed should be harvested from sound plants.

The soil should not have carried a diseased onion or shallot crop in the previous four years. Remains of a diseased onion crop should be burnt. When nearly all the onion tops in one onion-growing district were burnt for several years in succession, to control onion fly, the incidence of downy mildew was reduced at the same time.

Onion beds should be well drained and exposed to sun and wind. Spraying and dusting are less effective than one might desire. It is difficult to get the liquid or powder to adhere to the onion leaves. If it is tried a spreading agent should be added.

WHITE ROT

The disease usually appears about the end of May or beginning of June, especially in autumn-sown crops. The older leaves turn yellow and fall over and the plants can be easily pulled up because their roots are rotten. Round the base of the bulb there is a soft, flurry, white mycelium, hence the name " mouldy nose" sometimes applied to the disease.

In later stages this mould becomes matted and innumerable tiny, round, black sclerotia, none bigger than a pin's head, develop in it. They are readily detached and left in the soil where they may remain alive for many years. All varieties of onion may be attacked, though some are more resistant than others.

The fungus also attacks shallot, leek, and garlic and may occur on the wild crow garlic so that land may be infected with it that has never carried an onion crop. The sclerotia may also be found in seed samples and spread in that way. Probably the most important means of spread, however, is by infected sets and transplants. After an outbreak of white rot, onions, shallots and leeks should not be grown on the same ground for at least 4 years, preferably not for 8 or 10 years. Infected plants should be burnt.

SHANKING

This disease of onion and shallot closely resembles shanking of tulip. The outward symptoms are yellowing and shrivelling of the leaves and softening of the base of the plant. Some or all of the roots are soft and watersoaked. When-an affected bulb is cut longitudinally the bases of the youngest leaves and the tissues of the growing point are found to be discoloured, soft and often contracted.

The light-green or white healthy tissue contrasts sharply with the dull buff infected area. In some cases infection has been shown to occur from the soil in which the seedlings were raised prior to transplanting.

This is not always the case, however, as the disease may occur hi onions drilled out of doors. Onion seed should not be sown in soil that has carried a tomato crop, as the Phytophthora spp. which cause foot rot and damping off of tomatoes are the same as cause shanking. Onions for transplanting should be raised in sterilized soil.

ROOT ROT AND FOOT ROT

Various species of Fusarium may attack onion, both seedlings and older plants. Roots of seedlings often become attacked by Fusarium from infected seeds, and they turn yellowish-pink and decay. The plants put out new roots which are often also attacked.

The decay spreads upwards from the roots, and if the bulbs have formed the scales may rot at the base and the leaves turn yellow at the tip. The decayed parts of the bulb become reddish-brown or grey-brown, dry, mealy and overgrown by a pink or white mould; finally they become dry, shrivelled mummies. Alternatively there may be secondary infection by bacteria, causing a-soft, wet, stinking rot. Slightly infected bulbs may be overlooked at lifting and decay in storage.

MOULD ROT IN STORAGE

During storage onions may be seen with soft or dry areas on which there is a sparse blue-green covering of mould, usually beneath the outer scales. This is a species of Penicillium. A similar trouble occurs on garlic. The powdery black mould Aspergillus niger van Tiegh may occur in a similar situation and in Southern Europe a mould with yellowish-brown spore clusters causes rotting of garlic and onion bulbs. These moulds attack bruised bulbs or those damaged by frost or sunscald and stored in ill-ventilated places.

VIRUS DISEASES

These are common in shallot, much less so in onion, and are commonly referred to under the names yellow dwarf and mosaic, but the viruses causing them have not been extensively studied. Possibly some forms of the disease, however, may be caused by the common Cucumber Mosaic Virus, which has been shown to be transmitted by the shallot aphis, Myzus ascalonicus Doncaster.

Infected shallots are small with twisted, compressed, limp leaves, usually streaked with yellow. They make only small bulbs, which may be soft and not keep well. This condition is general where half a crop may be affected.

It is at least as common in English shallots and occurs to some extent in onions, especially in seed crops or in the vicinity of diseased shallots. Similar symptoms have been seen in the perennial Welsh onion, Allium cepa L. clon perutile Stearn. The shallot aphis occurs on stored bulbs as well as in the field, so possibly some transmission of the disease occurs in storage. Cucumber Mosaic Virus is also transmissible mechanically, as by bruising sets or transplants. Soil infection and seed-borne, infection have not been proved. Infections may not give rise to visible symptoms in the first year but the bulbs will yield diseased plants if replanted. Onion and shallot sets should only be taken from healthy crops.

If the health of a consignment is suspected a sample may be planted in a greenhouse during the winter, if infected bulbs are present they will give obviously diseased foliage. It was apparently this disease that Buchanan cured by keeping the bulbs at 84°F, for 112 hours.

TURNIP AND SWEDE

BROWN HEART

This disease, known as "Raan" in parts Brown heart tends to be worst and by a name meaning roots. It is caused by a deficiency of available "marbling "affects only the boron in the soil. In some districts, notably flesh of the root. The soils are often halfway between the centre and the rind is truly deficient in boron, but in other districts seen to be marked with irregular discoloured brown heart may be the result of temporary boron deficiency brought about by excessive liming.

In such cases use of an acid fertilizer like sulphate of ammonia in place of sodium or potassium nitrate may have a beneficial effect. Brown heart can be easily and cheaply prevented by supplying boron to the soil, preferably before the seed is sown, in the form of borax or boric acid. Because boron is poisonous to plants when present to excess the dressing should not exceed 20 Ib. borax or 15 Ib. boric-acid per acre, or 1 oz. to about 15 and 20 sq. yards respectively, and must be thoroughly and evenly mixed with the soil. It may also be applied mixed with the fertilizer, though preferably not with sulphate of ammonia.

PEA

LEAF SPOT, POD SPOT, AND FOOT ROT

The common leaf and pod spot fungus Ascochyta pisi causes small yellow brown spots with brown edges, dotted with small brown pycnidia, on leaves, stalks and pods. The attack is general and may be destructive. Very similar fungi occur on sweet pea and on broad bean but it are not known whether these can infect peas. If the fungus attacks the fruit stalk the pod falls.

From the infected pod the fungus often grows into the seeds. Hence the disease is commonly seed-borne. Infected seed may rot or may give rise to a seedling bearing diseased spots on the stem and the tip of the young shoot. These may die before emerging above ground or may survive and suffer from root rot, depending largely on the soil temperature and speed of germination. Two closely related fungi, Mycosphaerella pinodes Vestergr, which has an imperfect stage Ascochyta pinodes Jones, and A. pinodella L. K. Jones, also cause spotting of pea leaves and pods and are important causes of foot rot.

The leaf spots of M. pinodes are small, dark purplish brown, often with concentric rings. On the stems it causes purplish streaks and on the pods small purplish spots which may combine to form large blotches. A. pinodella causes small spots like M. pinodes but less numerous.

ROOT ROT AND FUSARIUM WILT

Root rot may sometimes be due to seed-borne fungi, but usually a poor seed bed and too frequent cropping with peas are important contributory factors. Repeated cropping with peas may succeed for a few years, but the only safe method of cropping is a long rotation to prevent the build-up of soil infection. Once soil has become heavily contaminated with root-rotting fungi it is difficult to get rid of them and several of the fungi concerned are not particular about the plants they attack. The best plan is to retain the name root rot for a disease in which either the roots or foot of the plant are attacked by a decay starting from the outside and turn brown or black.

Fusarium wilt (Fusarium oxysporum Schlecht.) is a distinct disease, in which there is no discolouration of the cortical tissues in roots or stem base but the vascular strand turns brown, the plants become yellow, either individually or in patches, and in severe infections they wilt and die. Affected plants yield very little. It is encouraged by heat and is therefore worst in warm climates. Root rot and Fusarium wilt not only cause loss of crop but lead to difficulties in canning peas, because the peas ripen irregularly and also tend to be abnormal in composition and hard when canned.

Seed disinfection may be helpful in certain cases, but the chief emphasis is to be laid on maintaining healthy, fertile soil, including the practice of a long rotation of crops. In Wisconsin addition of a complete artificial fertilizer to the seed drill has led to quicker germination and less root rot. Excess nitrogen and phosphoric acid makes peas more susceptible, while excess potassium makes them fairly resistant to Ascochyta and Mycosphaerella. Ensiling pea straw is insufficient to destroy root-rotting fungi on it; the straw from a diseased crop should be burnt.

DOWNY MILDEW

On the leaves of field and garden peas this fungus causes yellowish blotches covered on their lower surface by a whitish or pale-mauve layer of conidiophores and conidia. The infected areas may eventually turn brown and in some cases the whole plant may be permeated by the fungus.

The pods may be attacked and bear pale-green or light-brown spots. These seldom produce conidia but the resting bodies, oospores, of the fungus are formed within the tissues of the pod. Infected pods often become spotted and deformed but it is not possible to tell whether infection of the seed has taken place. The fungus overwinters in the soil and can infect pea seedlings as long as the soil remains cool, with a temperature below about 65°F. Spraying with Bordeaux mixture plus a spreader has been suggested as a means of control but must be carried out in the early stages of the disease if it is to be effective. Rotation of crops will help to keep down infection from the soil.

MILDEW

The race of this fungus found on peas does not attack

other plants and hence Continental authors often treat it as a distinct species E. Pisi DC. It causes the usual white powdery mildew on both sides of pea leaves, formed by the superficial mycelium of the fungus and its chains of conidia. The tiny brown perithecia are much less often seen. In early and severe infections the mildew may occur also on the pods. As a rule pea mildew, though conspicuous, causes little damage. In early infections dusting with sulphur or spraying with lime-sulphur may be tried. Severely infected straw should be burnt or ensiled. Resistance to mildew exists, based on several inherited factors, so that resistant varieties could no doubt be bred.

MARSH SPOT (MANGANESE DEFICIENCY)

In samples of pea seed, especially in that grown on wet soils, one often finds a more or less circular brown depression in the centre of the inner face of each cotyledon. This occurs especially in varieties with large wrinkled seed but there is no sign of disease on the seedcoat, or until the pea is cut across. Often, too, the tip of the tiny shoot, the plumule, is damaged, or when an affected pea germinates the tip of the shoot may die when only a few inches high.

Usually secondary lateral shoots develop to replace the dead tip, and if the seed is sown in fertile soil the plant ultimately suffers little from marsh spot. The trouble has been shown to be due to a deficiency of manganese and may be worst when ripening is slow. It may be prevented by applying manganese sulphate to the seed crop at the rate of about 1 to 10 cwt. per acre. Spraying the foliage with manganese sulphate solution during flowering may be even more effective.

PEA POOR GERMINATION

Poor germination of peas may be due to pre-emergence blight caused by Ascochyta, or to rotting of the seed by various soil fungi, such as species of Fusarium, Botrytis, Pythium and or with one of the proprietary organo-mercury various bacteria, especially if the seed was of seed disinfectants. Protection against these rotting germinates badly when sown too early.

RHUBARB

RUST

The cluster-cup (aecidium) stage of this rust occurs on rhubarb leaves. It causes carmine-red spots on the under-side of which is the tiny whitish cup with delicately fringed edges. The same stage occurs on leaves of sorrel and dock. It is common in the vicinity of swamps and ditches but not elsewhere. The uredospore and teleutospore stages of this rust occur in summer and winter on the common reed (Phragmites communis) where they cause yellowish-brown and black oblong pustules respectively.

DOWNY MILDEW

It has been found to be widespread but does little damage and is therefore seldom noticed by rhubarb growers.

Discoloured spots appear on the leaves and eventually turn pale with a red margin. Their size varies greatly. The usual layer of downy mildew conidiophores appears on the lower surface. On young spots they may form a continuous mauve coating but on old spots they occur only as scattered white specks. If the disease seems to be spreading the leaves should be sprayed or dusted with a copper fungicide. As a general rule there seems no necessity for any control measures to be undertaken.

LEAF SPOT

This fungus causes circular brownish sharply denned spots on the leaves, with a fine whitish mould on their under surface. If it appears to be on the increase a fungicidal spray or dust may be applied. The disease is only occasionally reported in Britain.

CROWN ROT

It fairly often happens that rhubarb rots at the crown, either starting in the crown and working a little way up the leaf stalks starting in the crown and spreading down into the roots. The leaves of such plants turn red or wither completely.

Often the diseased tissues are covered with a layer of grey mould but it does not always follow that grey mould was the primary cause of the trouble.

Other fungi or bacteria may have been the primary parasites which opened the way for the Botrytis. Verticillium albo-atrum may also attack this crop, causing the plants gradually to wither and die. The trouble is usually ascribed to Bacterium rhaponticum Millard, which causes the roots to become soft and almost chocolate-coloured.

The problem is complicated by the activities of the eelworm Anguillulina dipsaci Kiihn which is commonly associated with Bacterium rhaponticum in the rotting crowns and is apparently able to cause decay of the tissues on its own account. When a new rhubarb bed is made the mother plants should be very carefully selected and no part of a plant with any sign of crown rot should be used for planting.

Also rhubarb should be replanted straight away after being dug up, not, as is so often done, left for some time on top of the soil.

CARROT

VIOLET ROOT ROT

The fungus has a reddish-purple mycelium which grows in a network of fine threads over the underground parts of diseased plants, which gradually collapse as the tissue beneath the mycelium decays. It may grow through the soil to the next plant and attack it also.

In dry weedy soil the fungus may persist in the same place for many years. Many different kinds of plants may be attacked. Potato, turnip, swede, sugar-beet, mangold, broad bean, red, white and Alsike clover, asparagus, carrot, celery, chicory, parsley, salsify, seakale, carnation, holly, and the weeds wild carrot, plantain and creeping buttercup. It is the most injurious root disease of sugar-beet and becomes a serious problem where carrot growing is practised intensively. Baudys and others consider that different races of the fungus exist and call that on carrot.

This view does not find general acceptance and it is wiser to consider the possibility of infection spreading from any of the above-mentioned crops to any other. The fungus persists in the soil for many years and little can be done to get rid of it once it has appeared in a field. Long rotations, avoiding root crops for a considerable number of years, offer the best hope of control. On a small scale it may possibly be killed by soaking the soil with corrosive sublimate solution at the rate of 2 oz. sublimate plus 18 oz. saltpetre to 22 gallons of water.

CARROT FUNGUS

The disease appears in late summer on one-year-old carrots as sunken grey spots on the upper part of the root, often forming a sunken ring round the rosette of leaves. Many blackish pimple-like pycnidia appear on the surface of the spots, from which there ooze out in damp weather red globules or tendrils, composed of innumerable conidia embedded in mucilage not sink in as much as in Rostrup's description. The spots became blackish-brown, covered with a short loose grey mycelium mingled with white mycelial tufts. Pycnidia were gradually formed in the grey patches. The fungus may spread freely during storage and may grow from infected roots into healthy ones and destroy a whole consignment.

Numerous other fungi (Stemphylium radicinum, Sclerotinia sderotiorumy Botrytis cinerea, Fusarium spp.), however, often cooperate in this decay. The fungus does most damage to the seed crop. When infected roots are planted out to seed, spots gradually appear on the stem and leaf stalks, especially at the axils of the leaves.

Often the spots are situated one above another and run together to form a streak down the stem because the mycelium tends to spread along the furrows. Infection becomes widespread by means of the conidia during the summer. When flowering commences the disease has usually advanced so far that the root decays, the stem withers, the leaves hang down and the plant dies without setting seed.

Seeds that do mature may also be infected and bear small black pycnidia. The disease was formerly much more important than at present. Commercial seed production is now usually from carrots that overwinter where they were sown and are less exposed to the disease than carrots taken up in autumn, stored over winter and replanted in spring.

BLACK ROT

Dry blackish spots appear on the leaves, leafstalk and stems of growing plants but the phase of the disease most often seen is a slow rather hard blackish rot on the stored root. It usually occurs on the neck but may also occur farther down the side of the carrot.

Under dry conditions the rot is slow and dry but under damp conditions it becomes a wet rot covered with a blackish-grey mould Fusarium sp. and other secondary fungi often appear on the decaying spots. In carrots planted for seed the tops may wither. The disease is seed-borne and kills seedlings by rotting the roots and hypocotyledonary stem.

DOWNY MILDEW

The upper surface of the leaf bears yellowish spots which later turn black and may cause the leaves to shrivel and fall. On the under-side of the spots is a loose white mould. This fungus is much more common on parsnip and races of it exist on many wild umbelliferous plants such as goutweed, hedge parsley and angelica. These should not be tolerated in and around gardens as, even if the race of mildew attacking them does not infect carrot, they may also harbour the carrot fly. An attack of downy mildew may be checked at the outset by spraying or dusting with a copper fungicide. Remains of infected plants should be burnt or otherwise destroyed in autumn.

CRACKED ROOTS

Boron deficiency has not been reported in English carrot crops. In pot experiments with carrots on moorland peat, however, 0delien in Norway found the roots very much cracked. Manuring with small quantities of boron compounds prevented the cracking. On slightly acid peat soils in Canada longitudinal cracking of carrots has similarly been controlled by manuring with 20 Ib. borax per acre. Carrots grown in sand cultures without boron also developed wide cracks, and leaves which turned yellow round the edges. Longitudinal cracking may also be due, however, to a sudden acceleration of growth when rain follows a dry spell.

ROOT ROT OF CELERY AND SCAB OF CELERIAC

These diseases are of general occurrence but seldom attain any real importance. In celeriac, brown smooth spots appear on the skin of the crowns and gradually develop a cracked, warted skin. In early and severe attacks development of the crowns is checked and their upper surface becomes covered with a brown or yellowish-brown, scabby skin which may crack, and the crowns do not keep well. This condition, too, is sometimes called "rust" but it has no connection with true rust caused by Puccinia apii.

In the scabby patches tiny dark pycnidia are formed and these may also be found on seedlings at the base of the leaf stalk and occasionally on the seeds and leaves. In severe attacks seedlings still in the seed box may develop damping off symptoms, i.e. they rot at the soil level and die. Celery is attacked by P. apiicola mainly at the base of the leaf stalks and the rot spreads thence into the crown and the roots. The decayed leaf stalks are bluish-green or black and the rotting roots are brown. Celery roots may also sometimes be attacked by soil fungi such as species of Pythium, causing a reddish root rot, Thielaviopsis basicola, causing a black root rot, and the ubiquitous Corticium solani.

HEART ROT AND OTHER BACTERIAL DISEASES

Bacterial soft rot involving the heart leaves and spreading up the leaf stalks and down into the stock is not uncommon in celery. The cause is probably the common soft-rotting bacterium. Damage by the fly larvae may begin the trouble and it is likely to become prevalent when there is no proper rotation of crops.

Danish experiments indicate that excess lime and boron

deficiency are another important cause of heart rot. A non-parasitic form of black heart is widespread in celery fields. It seems to follow periods of either drought or excessive soil moisture, especially if the crop is growing rapidly. A purely bacterial disease is that caused by Pseudomonas jaggeri (Jagger) Stapp. This organism causes small rusty-brown spots on the leaves. It can be controlled by dusting the seed beds weekly with copper lime dust and in the field crop by spraying with Bordeaux mixture.

MOSAIC AND OTHER VIRUS DISEASES

Bitterness

Celery and celeriac belong to the group of plants which profit by chlorine. Hence plentiful manuring with a chlorine-containing potassium fertilizer should give a good flavour and good keeping quality (high sugar percentage, low nitrogen content); the bitter taste comes from too high a nitrogen content. In celeriac boron deficiency results in hollow crowns with dark tissue lining the cavities. The crowns tend to turn black when cooked. In severe outbreaks there is only a thin layer of normal tissue surrounding a large hollow.

In some cases there is also a yellowing of the leaves, particularly of the heart leaves, which may die completely and the tops have a tendency to lie along the ground. In such plants there is usually a cavity in the heart extending down into the root stock, into which one can stick one's finger. In other cases there is no outward sign of the damage, which is only discovered when the crowns are cut.

The disease appears to be purely physiological in nature and to be connected with particular soil and weather conditions. It seems most serious where a good growing period has been followed by drought and then by a further good growing period.

Different strains of celeriac differ in susceptibility to heart rot. A high soil reaction clearly favours the trouble and it can be prevented by manuring with borax. At the experimental station at Hornum about 1 oz. of borax to 36 square yards entirely prevented the appearance of these hollow crowns on ground where untreated celeriac was severely affected. This kind of hollowness seems to have no connection with the small hollow so often found in celeriac as to be regarded as more or less normal. This latter is often no larger than a hazel nut or a walnut and the surrounding tissue is quite white, though it may be a little soft around the cavity.

The size of the cavity may vary somewhat with variety and conditions of growth but is much less than that due to boron deficiency with its unsightly dark walls. With ordinary celery in Florida, boron deficiency has been found to cause browning of the heart leaves and the appearing of brown cracks and lines all over the ribs of the outer leaf stalks.

Rusty Markings in Celeriac

The swollen crowns of celeriac are sometimes stained by reddish markings either in winding lines like worm tracks or in flame-like shapes. This red colouration is often, mistakenly, called " rust," but it has nothing to do with the rust fungus. It is therefore best to restrict the name celery rust to the disease caused by Puccinia apii. Unsatisfactory manuring and sour soil may co-operate to produce these golden-reddish patches in the crowns. Very close planting has the same effect. Apparently also other but in recent experiments no clear differences factors, as yet undetermined, have some in- have been demonstrated in the behaviour of fluence on it. celeriac grown with different manurial treatments.

CHERVIL

Foot Rot

Under unfavourable circumstances young chervil plants may develop a constriction at the soil level due to infection by one of the damping off fungi, fall over and die. The lower surface of chervil leaves may become quite white with a layer of coni-diophores of this fungus. The corresponding parts of the upper surface show as yellow spots which later turn black and may cause the leaves to wither. Races of this fungus attack many cultivated and wild umbelliferous plants. Weeds belonging to this family should, therefore, be destroyed in and around the garden. In the early stages of an outbreak an attempt may be made to stop it by spraying with a copper fungicide. The remains of an infected crop should be burnt or otherwise destroyed in autumn.

SPINACH

ANTHRACNOSE

This fungus causes olive-green angular leaf spots that may turn yellow or almost white and dry up. They may join together especially along the leaf edges and cause extensive withering. Colletotrichum spinaciae seems to be as yet unknown but was found on spinach. It is not uncommon on spinach seed and may be very injurious to germination. Seed-borne infection up to 60 per cent of a sample has been found. Fungicidal treatment by shaking up the seed with an organo-mercury seed disinfectant should be tried. It may also be useful to spray the seed crop with Bordeaux mixture or similar fungicide.

LEAF SPOT

The early stages of attack consist of small, rounded, white spots, scattered over spinach leaves, mainly over the apical half. In later stages the spots are more numerous and may join up and are more or less covered with a fine olive-green or blackish mould. The first spots are usually on the upper surface of the leaf but in later phases of the disease the fungus is equally abundant on both surfaces. No control measures have been suggested.

DOWNY MILDEW

Irregular, ill-defined yellowish patches appear on the upper surface of the leaves, while on the corresponding parts of the under-surface there is a close, downy, grey mould that may acquire a lilac hue. Severely infected leaves are unsaleable. The fungus may overwinter as oospores in the soil and these have also been found mixed with the seed. A high degree of humidity, at least 85 per cent, is needed for production of conidia and these germinate only in the presence of surface moisture. Infection may occur at temperatures between 40° and 75°F with an optimum at about 48°F. At temperatures of 60-65°F infection may take place within 3 hours of inoculation. Spinach should not be grown on land that has already carried a diseased spinach crop and it is advisable to sow the spring crop at some distance from over-wintered spinach.

With indoor crops the temperature should be kept fairly high. On seedlings one may try spraying or dusting with a copper fungicide, taking care to cover both sides of the leaf. This is impracticable with older plants whose leaves are ready for table.

VIRUS DISEASES

Spinach Blight

This common virus causes a very destructive disease of spinach. Young leaves become yellowish, thickened, brittle and remain small with the edges inrolled and wrinkled. In severe cases some leaves may be reduced to a fine ribbon and others twisted, distorted and mottled. The outer leaves later turn yellow, then brown and shrivel until only a central cluster of yellow malformed leaves is left. When seedlings are attacked they become completely yellow or die. The virus may be transmitted mechanically with the sap or, as occurs in nature, by aphides.

Mosaic

Mosaic of spinach caused by infection with the common beet mosaic virus from beetroot or sugar-beet is well known and has been produced experimentally. The first sign of infection is a downward arching of the young leaves, accompanied by the appearance of numerous bright golden-yellow spots. The yellow colour gradually spreads, and then the outer leaves die back beginning at the tip. The virus is spread by the beet aphis and other species and can be transmitted artificially by sap inoculation.

The incubation period is 1-3 weeks. White goosefoot or fat hens (Chenopodiwn album) may also be infected by this virus. It develops isolated yellow spots and the leaf edges curl under. The sow thistle may also become infected. These weed hosts may play a part in enabling the virus to survive between beet and spinach crops.

STRAWBERRY

MILDEW

Symptoms: Mildew may attack strawberries out of doors or under glass. When the leaves are attacked they bear only a thin coating of the floury white spores among the hairs on their under surface. On the corresponding parts of the upper surface dark purplish blotches appear and the leaves roll upwards, exposing their under surface. On the fruit mildew produces a similar sparse white powdery coating and if they are infected before they ripen them do not turn a good red colour but-become dry and poorly coloured. Their market value is thus impaired.

Wide variations in temperature appear to encourage the disease and should be avoided under glass and in frames. The usual sulphur fungicides may be applied until the fruit begins to ripen and after it has been picked.

LEAF SPOT

Reddish-purple spots appear on the leaves which become more or less circular and turn white at the centre, which may die and fall out. In severe infections the spots may be small and angular, but crowded together so that large parts of the leaves turn brown and shrivel. During the summer a white mould composed of bunches of short hypha: bearing conidia of the Ramularia stage appears on the centre of the spots. On over-wintered leaves the tiny black Mycosphaerella perithecia may develop, sunken in the leaf tissue. They are, however, seldom seen.

LEAF BLOTCH

This disease starts as small purple spots like those of leaf spot but never develops the white centre so characteristic of that disease. The spots gradually increase in size up to an inch or more and may involve an entire leaflet. The large spots are dark-brown, often paler at the centre, with a dark-purplish edge, sometimes fringed by a zone of yellowish leaf tissue. On the spots there appear minute brown pycnidia and in artificial culture a perithecial ascus stage has been obtained. This belongs to the genus Gnomonia.

LEAF SCORCH

This disease starts as small dark purplish spots like those of leaf spot. As they increase in size, however, they do not develop the white centre. Individual spots may unite to cover a large part of the leaflet and turn dark brown, hence the name " scorch." Infected leaves often turn yellow and the outer ones may wither altogether while the central leaves have few or no spots. Spots may also occur on the fruit stalks and check development of the berries. Tiny black conidial fructifications of the fungus develop on the upper surface of the spots.

Somewhat similar leaf spots, brown with a reddish-purple border and often a yellowish zone outside that, may be caused by another pycnidial fungus, Septogloeum fragariae Hohnel, which also causes hard rot of the fruit.

GREY MOULD

Many berries can be spoilt and covered by this hairy grey mould. Excessive foliage favours the attack, for which reason it is advisable to avoid heavy manuring with nitrogen. The second and third years' crops are most likely to suffer. On low wet ground the fungus may also attack and kill the root stock and destroy the whole plant.

On plants under glass all mouldy berries and any diseased or dying leaves should be picked off. Supporting the fruit on short forked twigs helps it to escape infection. Preventive spraying with Pomarsol has been tried. Out of doors one should avoid too close planting and should spread straw or chaff under the trusses after flowering. The variety Deutsch Evern is said to be particularly susceptible to grey mould.

VERTICILLIUM WILT

Sometimes the outer leaves wither first, turning brown and dry, or develop a brown margin. In other cases the whole plant withers at once and dies quickly. In milder attacks it may linger for a long time and develop a few narrow leaves and weak runners that soon die. Usually there is no discolouration in the root stock until the disease is fairly well advanced, when small dark spots can be found in the flesh. Later on the roots turn dark brown. One can isolate the fungus from the roots much less readily than from the root stock. Usually one can obtain it from the upper part of the root stock and even in the petiole.

Verticillium wilt is not the usual cause of black roots. Runners should be taken only from healthy plants and not from those adjacent to wilted ones. It is, indeed, best not to take any runners from beds in which disease is present. Strawberries should not be planted on land known or suspected to contain Verticillium from another susceptible crop such as potato, tomato, raspberry or stone fruit.

LEATHERY ROT

In this disease the developing fruit turns brown and, instead of becoming soft and rotten as with grey mould, it becomes tough, leathery and eventually shrivels. As the rot proceeds there is a sharp dividing line between healthy and diseased parts of the berry.

The fungus may attack a great many host plants and there is no reason to suppose the late date of its discovery indicates it to have been a recent introduction. Heavy rain favours the disease. Infection comes from the soil and is no doubt spread by splashing during rain.

It is therefore apt to be more severe when the beds have not been properly strawed. A copper lime dust applied 5-6 weeks prior to ripening, followed by a second and third, the last very light to avoid poisoning the fruit, gave best control, but a sulphur dust was also beneficial and less objectionable for use on a soft-fruit crop.

"RUNNING OUT" OR "DECLINE" OF STRAWBERRY CROPS

It is a common experience that after a time strawberries cease to " do" at a particular place and the soil is then often said to have become " tired " of the crop. As in the case of other " tired soils " the phenomenon is not a simple one but may be due to several causes, operating singly or in combination.

In the case of strawberries these include cultural factors, virus diseases, root rots, red core disease, and insect pests. Strawberry plants suffering from several diseases and pests tend to develop much the same type of "degeneration," small, weak plants with short-stalked, red-edged leaves, few runners and low cropping capacity. The roots may be damaged by beetle larvae, especially the June. Where these insects are plentiful the plants wilt and die, especially in June and July. Nematode infection produces the disease known as red plant.

The mites infest the leaves which remain small, brittle and yellowish and in extreme cases turn red. Not only do they weaken the plant but the damage they cause masks the symptoms of virus diseases.

Soil conditions in many places are unsuitable for strawberry cultivation. Very heavy soils should be avoided, so should very light sands, except in areas of high rainfall. Strawberries require humus, and a direct relationship may be traced between the organic-matter content of the soil and the vigour of the plants. Mammal experiments have often shown good results from the use of stable manure.

At Long Ashton manuring with waste products of the woollen industry has led to production of high-quality fruit. If a heavy dressing of farmyard manure has been applied before planting, application of artificial manures seems superfluous, though a little nitrate may be given at the time of blossoming and a little more plus superphosphate after the crop has been picked.

When dung is not available strawberries respond to phosphatic fertilizers, especially bone meal or superphosphate, plus a small quantity of nitrate of soda or sulphate of ammonia. Phosphorus-deficient strawberries develop purple and bronze tints in their leaves. Heavy potash manuring is not required except on definitely potassium-deficient soils. If there is chlorine in the fertilizer it may be definitely harmful, so too is sodium.

Acid soils of pH 6-7 are said to be best but experience at East Mailing shows that strawberries may do well on soil rich in lime. Presence of lime may, however, lead to minor element deficiencies and excellent results have been obtained by micro-nutrient manuring on limed soils. It may be necessary to supply manganese, iron and copper; Boron deficiency has been demonstrated in sand-culture experiments but not yet in the field. Methods of cultivation too are not always suited to the requirements of the strawberry plant. Its new roots are formed mainly in late summer and autumn and arise higher up on the root stock than the old ones.

If the natural growing conditions of the plant are to be reproduced it should be surrounded with a layer of dead leaves or similar material every year. At Long Ashton good results have followed surface hoeing in August, after which the soil was pressed firmly against the plants.

The practice of allowing weeds to grow in the beds after the fruit has been picked and digging them in in autumn is undesirable as the digging destroys the new roots of the strawberries as well as the weeds. Runners take much nourishment from the parent plant and should be removed from a fruiting bed. Runner production should be carried on in separate beds and preferably as a separate industry. It is usually best to pick the blossoms from plants in their first year after planting and to allow them to fruit for the first time in the second year.

A plentiful supply of moisture is necessary for a good crop of fruit and in dry summers irrigation may be a profitable practice. On the other hand ill-drained soils are unsuitable and encourage root rots and red-core disease. Frost may damage many plants, and low-lying areas should not be chosen for strawberry cultivation. The practice of covering the beds with straw helps to protect against frost and drought.

To protect the blossoms from late frosts straw should be placed above the plant at night and raked away during the day. It should not be strewn on the ground below the blossoms until frost is no longer to be feared or frost damage may be intensified. Brown roots.

At the lower end of the root stock on any strawberry plant there are always some of last year's roots that have turned blackish-brown. The new roots that arise mainly in autumn higher up on the root stock and from the runners should be white or yellowish throughout, not dark brown or black. A host of fungi is to be found on and in the brown and black roots, many of which are common inhabitants of diseased roots in many plants.

Some of them have been shown to be capable of attacking healthy strawberry roots under certain soil conditions. The presence and extent of brown root rot in strawberries depends on the inter-relationship of fungi and soil conditions. In this connection we may mention Hildebrand and Koch's experiments in Canada with the " Wisconsin Tank," an apparatus in which the plants can be grown at a number of controlled temperatures and amounts of soil moisture. In this way it was demonstrated that there appeared to be no simple relationship between fungal activity and major nutrients, minor nutrients or soil reaction, but that soil moisture and temperature exercised an important effect.

At a soil temperature of 45-50°F and at 60 per cent soil water-holding capacity the strawberry roots remained quite white, but as the temperature rose to 60°F and moisture content to 80 per cent browning of the roots steadily increased. Brown root rot of strawberries is to be controlled mainly by maintaining good growing conditions with a well-drained soil. Only runners with clean healthy roots should be selected for propagation. In laying out new strawberry beds it is desirable to avoid land that has carried either strawberries or raspberries within the last 4-5 years. Planting in early autumn encourages strong growth.

RED CORE

This is a serious root disease caused by Phytophthora fragariae Hickman. The finer rootlets are first destroyed so that the main roots alone remain, looking like "rats' tails" when the root system is washed free from soil. These main roots are then attacked at the tip and the fungus spreads upwards through the vascular cylinder, which it stains reddish-brown, the "red core." The root cortex remains white and healthy for a time but eventually turns black and decays. The fungus has been isolated in pure culture who was able to reproduce the disease by inoculating P.fragariae to healthy plants.

Red core is worst on soils with high water content but cannot be controlled by draining infected land. It is probably spread largely by the sale of infected runners and perhaps locally by infected drainage water as the fungus belongs to the group commonly called water moulds. Research sponsored by the Department of Agriculture for Scotland has led to the breeding and selection of a number of highly resistant strawberry varieties of good commercial quality.

These are known as the Auchin-cruive seedlings, from the place where they were bred, the different varieties being distinguished by numbers. No method of soil treatment has proved effective against red-core disease, and the growth of resistant varieties offers the only hope of successful strawberry cultivation on infected land.

CAULIFLOWER

FINGER AND TOE

In cabbage and its allies the spindle-shaped irregular swellings on the roots are the most obvious symptoms, and are often accompanied by poor development of the head and, in dry weather, by flagging and a bluish tint to the foliage. In swedes, turnips and radishes, where the stem swells to form storage of reserve foodstuff, this is also attacked, and becomes covered with tumours and cracks both above and below soil level. In the swelling the cells of the flesh are enormously enlarged and developed at the expense of the woody tissue so that the tumours easily rot, the decay being completed by bacterial activity. The galls caused on roots by the cabbage gall weevil can easily be distinguished by their more spherical shape and by always having a cavity at the centre, containing the larva.

The so-called hybridization nodules on swede roots are also more round, have a more wrinkled surface and never become very large. True club root can always be recognized under the microscope by the gigantic malformed cells of the host. These cells contain innumerable minute spores of the fungus which are liberated into the soil when the swellings decay. They may remain alive in soil for at least 5-6 years. Living roots of cruciferous plants stimulate the spores to germinate and from each emerges a swarm spore, a naked.speck of protoplasm able to creep and swim through the soil moisture to the root.

It may penetrate the latter through wounds or cracks, as where side roots have emerged, but often swarm spores perforate the wall of root hairs and grow and divide within the hair cell. From the hair the fungus protoplasm penetrates deeper into the root, following especially the cells of the medullary rays between the vascular strands. These thin-walled cells are stimulated to enormous development and cause distortion and disruption of the whole tissue. Within the giant cells many swarm spores unite to form a large mass of fungus protoplasm (plasmodium) which ultimately breaks up into hundreds of tiny resting spores, ready to be set free in the soil by decay of the root.

CLUB ROOT

Infection can take place at temperatures between 43° and 81°F, but from 64-77°F is the most favourable temperature. This is why late sowings generally suffer much more than early sowings made while the soil is cold. Soil reaction plays a very important part; pH 5.8 is most favourable for spore germination. Swarm spores can emerge from the resting spores at pH 7.5, but are then sluggish. When the soil is packed with spores, however, infection may occur even at pH 8.3. Similarly, experiments have been found that club root infection is distinctly diminished by increasing dressings of lime or marl, but that severe infestations cannot be stopped by liming alone.

Sometimes even slight differences in soil reaction have an effect and nitro-chalk gives a healthy crop where sulphate of ammonia gives a diseased one. Sodium, magnesium and barium salts that raise the pH value in the same way as lime have a corresponding effect on the fungus. Soil moisture is essential for club root infection; if the soil contains only 50 per cent of the water it is able to hold, cabbage can still grow in it, but the swarm spores are unable to move; as the water content rises above this level so does the percentage infection rise in proportion. High soil moisture during winter and heavy spring rains lead to virulent outbreaks.

Summer droughts make the plants flag conspicuously so that they are weakened, but infection by the fungus does not increase. Plants from seed sown in August may escape infection in the autumn, but in the following spring, when new roots are produced, these are liable to be attacked. Heavy summer rains may speed up rotting of roots that are already attacked. Club root does not spread in stored produce but if conditions are moist the diseased tissues quickly decay. The sites of turnip should never be used for cultivation of a susceptible crop.

The type of soil is not very important, except that well-drained soils rich in lime are least prone to club root.

Spread of infection. Soil on which a crop infected with club root has been grown may be regarded as infected for at least 5-6 years. If a cruciferous crop is taken before this time has elapsed, or if many cruciferous ground-keepers and weeds occur, the period of soil infection is correspondingly extended.

The sale of soil to gardens, where the purchaser did not know it to be contaminated with club root, judgment being given for the purchaser, so that the seller was obliged to make good the damaged crop.

Soil is easily spilt in transport and infection may thus be spread to other areas. Infected material may also be carried by tools, on horses' hoofs, tractor treads and the boots of workers. It seems that it may be also spread to hitherto uninfected areas by particles of soil blown by the wind. Infection may also be water-borne, as when diseased turnips have been thrown into a pit which overflowed as a result of heavy rains. Infection in farmyard manure is a very serious problem. Observations indicate that the spores may pass through the intestines of ruminants in a living condition, but this is probably not of practical importance.

The damage is done by contamination of stable manure with waste infected fodder such as turnips. The sediment of liquid manure may also be infectious. Hence bought stable manure and kitchen refuse are to be mistrusted, but on the whole consistent use of heavy dressings of non-infected stable manure seems to lead to less club root. Cabbage plants frequently become infected in the seed bed, and are sold when bearing only a slight swelling easily overlooked. These are a frequent source of infection in private gardens. The seed itself cannot be infected, but infection may be present in soil particles in a seed sample.

CANKER

Phoma lingam may attack seedlings, half-grown plants, cabbage heads in storage, seed-bearing plants and the seed itself. It is one of the causes of damping off in cabbage, broccoli, and turnip seedlings. The plants rot at soil level, and when the disease was caused by P. lingam small dark pycnidia appear on them, out of which, under moist conditions, ooze amethyst-coloured drops containing the spores P. lingam infection at this stage comes from infected seed. These seedlings attacked by the primary seed-borne phase of the disease almost always die, but the disease spreads from them to other plants in the seed-bed. These when planted out develop the canker phase. In this the stem is usually invaded just below soil level.

It turns black and decays so that the plant becomes loose in the soil and may eventually fall over. Secondary roots may be produced from the stem above the canker and keep the plant alive for some time. The leaves often take on a blue or purplish hue, but similar colouration may result from other injuries to the root system as by larvae of the cabbage fly.

Pycnidia form on the rotting tissue or at the edge of the canker next to the healthy stem. The disease may progress further in cabbage heads stored for the winter. The fungus grows up through the stump into the leaves, which turn black from the base upwards and outwards. They become loosened from the stem so as to produce a curious crater-like hollow round it. Ultimately pycnidia appear in the diseased tissue.

If there are a few infected plants in a seed crop the disease may easily spread considerably on the ripening pods.

Phoma lingam infection is often confused with " dark spot" caused by Alternaria brassicae and A. oleracea, but it is of great importance because it leads to production of infected seed. In a seeding crop numerous dark spots appear on leaves, stems and pods and on these the pycnidia are gradually produced. From the spots on the pods the fungus penetrates the seeds where it occurs as resting mycelium between the embryo and the seed coat.

Thus infection of the seed takes place between flowering and seed ripening, but the seedling becomes infected during germination. It is not possible to tell whether a seed sample is infected or not merely by inspecting it, though some infected seeds may be lighter, greyish and somewhat shrivelled.

These would float if the sample were shaken up in water. If a germination test is made, however, the diseased seeds can be detected by the pycnidia which appear on them and on the seedlings grown from them. Such a test takes from 3 to 4 weeks. The same fungus causes a disease of swedes called dry rot. On that host it has been named Phoma napobrassicae Rostr. Numerous other names have been given to it, such as P. oleracea Sacc., P. brassicae Thiim, P. siliquastrum Desm. and Aposphaeria brassicae Thiim., but they are all superseded by the oldest, viz. Phoma lingam (Tode) Desm.

RING SPOT

Almost all above-ground parts of the plant may be affected. On the leaves the spots are usually numerous, round, and light brown to grey-brown with a narrow olive-green border. They bear innumerable small pycnidia or perithecia, usually arranged in concentric rings, whence the name ring spot. The fructifications may be so numerous that the whole spot looks black. In severe attacks the outer leaves of cabbage turn yellow and fall.

Spots on the outer leaves of a cabbage head may be so numerous that from a short distance it resembles a red cabbage. Similar spots occur on stems and pods of plants grown for seed. Whereas pods spotted by Alternaria tend to open prematurely, those attacked by ring spot are apt to remain closed even when the seed is ripe.

Phoma lingam, the cause of cabbage canker, may be responsible for somewhat similar spots, but there is no connection between it and the true ring spot fungus. P. lingam spots bear far fewer pycnidia.

The ring spot perithecia form during autumn and ripen about December. Recent investigations have shown that the spores produced by the pycnidia are not conidia and are unable to infect leaves or to grow in artificial culture. resumably they are male cells or spermatia, analogous to those produced by rust fungi, but their precise function remains unknown. Ascospore germination is slow at 36°F, but rapid at temperatures between 60° and 72°F.

At the latter temperature infection may take place in about a day, but there is an incubation period of 2 to 3 weeks before the spots are clearly developed. The fungus is believed to survive from one crop to the next on old leaves, stems and seed pods. Infection is mainly by ascospores blown by the wind.

DOWNY MILDEW

The upper surface of the leaf bears small, pale-yellow, angular spots, often speckled with brown, which may grow together to form irregular patches. On the under-surface the corresponding areas are covered with a light-grey mould formed by the much-branched conidiophores and conidia of the fungus. The disease is commonest on seedlings in the seed bed and on the lower leaves of young plants, but it may occur on older plants and even on stored cabbage heads. In the latter circumstances the infection came through the stalk from infected lower leaves of the mature cabbage.

The pith turns dark grey, and grey lesions with a yellow-brown margin occur on the leaves throughout the head. Downy mildew may also spoil the sprouts in brussels sprouts. The downy mildew affecting most cruciferous plants, cabbage and its allies, turnip, swede, seakale, mustard, radish, stock, wallflower and weeds like shepherd's purse and bitter cress, has traditionally been regarded as a single species, Peronospora parasitica Tul.

This name is still used for it in British literature. In Switzerland, however, Gaumann has shown that P. parasitica in this wide sense includes a considerable number of races specialized on particular host plants and these he prefers to call different species.

In his nomenclature the downy mildew of cabbage and its allies becomes P. brassicae Gaumann. Because of this degree of specialization there is no risk of cabbage being infected from, the downy mildew so common on shepherd's purse. Within the tissue of the diseased leaves the fungus forms its sexual reproductive bodies, oospores, which can remain alive in the decaying leaves in soil for several years.

They are, however, not very abundantly produced and the fungus probably overwinters mostly in the form of mycelium in living plants. Laubert mentions the possibility of the mycelium growing into the seed pods and seeds, but seed-borne infection is not usually regarded as important in this disease.

The fungus can grow and form conidia at a temperature as low as 40°F and the spores germinate readily in water between 46° and 54°F, less freely at 66°F and scarcely at all at 73-77°F. Importance.

Downy mildew does most damage to plants ir loors and in frames. Only occasionally under moist conditions does it become important out of doors. Plants infected in the seedling stage usually grow away from the disease. It has damaged heads of savoys particularly by persisting through the summer on the lower leaves, attacking the outer leaves of the head and growing in from there to cause brown spotting of the inner leaves in stored heads. Kale seems less susceptible than cabbage. Soil in which cabbages or their allies have been grown during the last few years should not be used for another cabbage crop.

Development of the fungus is checked by dry air, hence the plants should be kept dry and the atmospheric humidity low by giving plenty of ventilation to seed beds in frames and glasshouses. Spraying with Bordeaux mixture and other copper sprays also definitely checks downy mildew, especially when the under-side of the leaves is wetted, but spraying should be done early in the day and in weather when the moisture will dry quickly. Copper lime dusts may also be helpful, sulphur dusts are less effective.

WHITE BLISTER

This disease is also sometimes called white rust, but that is a misleading name as the fungus which causes it is allied to the downy mildews and not to the true rust fungi. It is found on almost all members of the family Cructferae: turnip, swede, cabbage, and its allies, mustard, cress, horseradish, arabis, honesty, wallflower, and numerous weeds, especially shepherd's purse. Any above-ground part of the plant may be attacked and in flowering plants the stems are often twisted, contorted and much swollen. The spores are formed in large white blisters under the epidermis, which burst open to expose powdery white pustules.

DARK LEAF SPOT

The dark leaf spot fungi are commonly seed-borne and may attack seedlings, causing brown spots and streaks on stems and leaves. This phase of the attack is of minor importance in western Europe, but spotting of mature leaves and especially of the flower stalks, flowers and pods of seed crops may be of some significance. Two related fungi are concerned in the disease, which differ mainly by the size of their spores and other microscopic characters. Both cause rounded spots up to an inch across on leaves. The spots are pale yellow with a yellow margin on green leaves and green on red cabbage leaves, best seen on the upper side of the leaf. In damp weather the upper surface, and to some extent the lower also, becomes covered by conidiophores and conidia.

In A. brassicae they are pear-shaped with long beaks and form a pale-greyish felt-like coating; in A. oleracea there is a dark olive-green, closely velvety growth, bearing the conidia in concentric rings.

The spots can be found on leaves all the year round. In seed crops, when the weather is damp and warm, these fungi spread rapidly over the stems after flowering and form pale oblong spots, soon covered with a dark layer of conidia, on the ripening pods. In a few days of favourable weather the appearance of a whole seed bed may be changed from green to brown or almost black. On the pods the mycelium grows through the wall into the seeds.

Severely attacked pods shrivel and die. In others the seeds ripen but are weak and germinate with only a weak shoot soon killed by the fungus. Slightly infected seeds can be protected by seed disinfection and will in any case give a vigourous seedling if conditions favour rapid germination. Presence of these moulds causes the pods to split open prematurely so that much seed is lost by shedding and the yield obtained from the crop is reduced. The quality of the seed is also adversely affected. Spots on the leaves are seldom of much importance in themselves, but help the fungi to persist in the crop all the year round.

The lowest leaves of a young seed crop are usually infected. They wither in the spring and on them are produced enormous numbers of conidia by which the flowering and fruiting stems become infected later on.

GREY MOULD

Cabbage heads, especially when stored, may be attacked by the common grey mould, which causes watersoaked grey-green to brownish patches on the leaves. These become covered with the familiar grey coating of conidiophores and conidia and eventually irregular black sclerotia form in the rotting tissue.

Grey mould thrives under damp conditions, especially at a moderate temperature, and may cause great loss to cabbages in storage. It is important that the cabbages be dry and well aired for some days before being stored, and that care be taken to avoid bruising them, as the fungus easily gains entry to the tissues through wounds.

Hearts of spring cabbage may also be attacked in the field, especially after damage by frost. Botrytis cinerea may also damage brussels sprouts even out of doors during the winter months. Early Dwarf and Long Island are particularly susceptible to it. Early varieties generally suffer most.

SCLEROTINIA ROT

In cabbage this is principally a disease of the stored heads. Larger or smaller parts of them turn black and slimy and on these the white, cottony mycelium characteristic of the fungus develops.

Associated with it are large sclerotia, at first white, then with a black rind. Cabbages should be dry when stored and should be handled carefully so as not to be bruised, for damaged tissue forms an easy means of access for this fungus and for Botrytis cinerea, etc. The disease is worst when the straw in the pit becomes damp.

Cabbage varieties differ in susceptibility to Sclerotinia rot; of red cabbages Amager and spring red cabbage are resistant, and of ordinary cabbages high and half-high Amager.

BACTERIAL LEAF SPOT

This disease, has been found on cabbage, cauliflower, broccoli, savoy and brussels sprouts. Numerous small watersoaked spots appear on the leaves and soon turn brownish or grey with translucent edges. At first they are round but become angular or so by the fine veins of the leaf. If they are very numerous several may coalesce and the leaves may become ragged and torn. In severe infections the leaves may turn yellow and fall. The causal bacterium enters leaves through the stomata and can also attack flowers and the curd of cauliflower heads. It grows best at 77°F, not at all at temperatures over 86°F and is favoured by high humidity. Seed disinfection with corrosive sublimate may help to prevent infection, and spraying with Bordeaux mixture plus a spreading agent may check its spread.

BLACK ROT

Black rot may affect almost any cultivated cruciferous crop and some wild species as well, especially cabbage, red cabbage, savoy, brussels sprouts, broccoli, cauliflower, kohlrabi, swede, turnip, mustard, charlock, radish, wallflower and stock. Plants may be attacked at any age but especially when nearing maturity.

In cabbage a larger or smaller section of a leaf typically turns yellow and bears a network of blackened veins. This blackening spreads down the vascular bundles, to the upper part of the stem. Especially in turnip, infection may also start in the roots, which have been damaged by larvae of the cabbage fly and other insects.

In severe attacks so many of the outer leaves are destroyed that the cabbage plant appears " leggy " and its head becomes lopsided and loose-leaved. The blackened vascular bundles may appear in the head leaves too.

When infection comes from the seed the cotyledons become infected 12-20 days after sowing. Many plants may be killed in the seed-bed or, if they survive, spread the disease when they are planted out. The rot caused by X. campestris alone is rather dry and limited in extent but it may open the way to secondary organisms which set up a rapid evil-smelling soft rot. Infection may come from the seed or the soil, especially if diseased cabbage or turnip heads, roots or leaves have come in with manure.

From diseased plants and from infected soil the bacteria are spread by rain splashes, also by snails, by the larva; of insects like the cabbage fly and by workers among the plants. Most infections occur through the water pores on the edge of the leaves.

These exude drops of water during the night and if bacteria get into these drops they multiply and enter the terminal strand of the vascular system below the water pore. The bacteria can grow at temperatures between 40 and 100°F.

They thrive on cabbage which has heated in a silo. Freezing and thawing does not kill the bacteria, which are able to remain alive in the soil from one year to the next. Nor are they killed by drying and survive for one year, possibly for three years, in stored seed. Black rot seems to occur throughout the world wherever members of the cabbage family are grown.

SOFT ROT

Plant organs that are used for the storage of foodstuffs are often affected by a soft stinking rot caused by bacteria. Probably the commonest cause of the trouble is Bacterium carotovorum which has received several other names in the past according to the host plant in which it was found, e.g. Bacillus oleraceae Harr., B. omnivorus Van Hall, B. apiovorus Worm, and B. hyacinthi septicus Heinz. Closely allied species of soft-rotting bacteria are Bacterium aroideae (Towns.) Stapp, which can attack arum corms, iris leaves, potato tubers, etc., and Bacterium phytophthorum (Appel) Burgwitz, the cause of black-leg of potato.

Host plants

Soft rot is most frequently found in mature turnip roots, next in carrots and quite often in radish, swede, cabbage and its allies, potato tubers, especially in wet, warm soil, celery, seakale, parsnip, onion, leek, cucumber, melon, horseradish, tomato, capsicum, egg plant, arum, iris, hyacinth, cyclamen, and probably parsley, wallflower and violet.

It is always soft and fleshy organs that are attacked, usually storage organs, like carrot and turnip roots, melon fruits, iris rhizomes, but sometimes other organs like the soft stems of cucumber and melon and the fleshy leaf bases of iris. It is characteristic of soft rot that the tissues dissolve into a sloshy mess unable to support their own weight. The rotting organ therefore collapses while watery drops and slime containing swarms of bacteria ooze out. There is often a peculiar smell such as one associate with rotten cabbage. Bacterium carotovorum is able to move freely in water by means of its flagellae.

It is emphatically a wound parasite but once it has obtained entrance to plant tissues it secretes a poison which diffuses into neighbouring healthy cells and kills them. The middle lamella, the substance between plant cells which binds them together, becomes dissolved, so that the tissue becomes a structureless mass.

Usually there is no change in colour but sometimes there is a slight brown discolouration. The bacteria multiply most rapidly at 80-86°F, hence their activity is restricted by cold storage. Rotting at the neck in swedes and turnips similarly often follows injury by the swede midge. The heart leaves of celery are often destroyed by this organism.

BROCCOLI MOSAIC

The first symptom of infection in broccoli is a clearing of the finer leaf veins which appears first towards the base of young leaves. It may become more obvious as the leaf develops so that the veins become quite yellow. This condition is slowly replaced by vein banding, in which a dark-green band of tissue follows the larger veins and the intervening portions of the leaf blade become pale or yellowish.

Small dead spots, at first white, then pale brown, may appear in the yellow areas. The leaves may be twisted and deformed and there is a tendency for the outer leaves to bend backwards and downwards instead of arching over the curd. In severe infections the whole plant is stunted and no marketable curd is produced.

CABBAGE BLACK RING SPOT

This disease is common on cabbage and is easily recognized by the small black ring-like markings which occur scattered over the leaves of infected plants. On cauliflower the virus causes a fine light-green mottle. The virus also attacks a number of ornamental plants such as arabis, wallflower, hesperis, and stock, and causes "breaking" in the colour of self-coloured flowers. It is transmitted by aphides.

BROWNING OF CAULIFLOWER

Boron deficiency on cauliflower is widespread in some districts. Small transparent areas appear in the pith of the stem and on the surface of the maturing head. These turn brown or even almost black. Aifected heads do not decay but are very bitter and unpalatable. In severe boron deficiency the leaves may be narrow and the formation of a head may be delayed or even prevented. The disease can be controlled by manuring with borax at the rate of 5 to 25 Ib. per acre according to the severity of the trouble. The usual symptoms are watersoaked spots in the pith of the main stem. These turn brown and may break down into cavities.

In extreme cases the oldest leaves may be mottled at the edges, be curled and irregular in shape and develop cross-cracking on the upper surface of the leaf stalk. Early maturing varieties are most prone to the trouble.

It seems that when potassium is deficient cabbage and all its allies develop pale leaf edges which later shrivel, also, sometimes dead spots on the leaf blade and brown streaks on the leaf stalks. The heads may be loose and defective, brussels sprouts develop few sprouts, there are green leaves in cauliflower heads and red cabbage leaves turn greenish. Potassium-deficient plants seem more susceptible to aphis and downy mildew than normal plants.

WHIPTAIL

Whiptail is a condition in which the leaves, usually of cauliflowers, are narrow, irregular and wavy and in extreme cases consist merely of a bare midrib or of a midrib with irregular lodes of leaf blade on each side of it. The condition has been reproduced artificially by growing cauliflower and broccoli plants in sand cultures lacking molybdenum.

Whiptail usually occurs on acid soils very deficient in lime, but liming alone does not always cure it. Molybdenum in the soil is believed to be more available to plants in alkaline or limed soils than in acid ones. The disease has been cured by manuring with very small quantities of sodium or ammonium molybdate.

BLACK SPOT OF CABBAGES

There were complaints from several places that some of the leaves in mature cabbage heads were covered with small black spots. The appearance of the diseased cells resembled that of cells in turnips affected with brown heart, and boron deficiency may have been involved in the cabbage trouble. Similar black specks in cabbage and cauliflower leaves may arise in transit and during storage.

They have been produced experimentally by keeping cabbages at room temperature in an atmosphere with little or no oxygen and by keeping them in a well-ventilated room at temperatures near 32°F for 2-3 weeks. Black spot has also been found in the field after a sharp fall in temperature

SPLITTING OF THE HEAD IN CABBAGE

In very wet weather the roots absorb so much water that growth of the inner juicy tissues of the cabbage head becomes too rapid and the outer tissues are burst open. This is especially likely to happen when heavy rain follows a drought. By driving a spade into the soil on both sides of the plant or by twisting it half round in the soil a considerable number of roots will be severed and excess absorption of water will be prevented.

MAIZE

SMUT

Maize smut may attack all the young parts of the plant above ground, even the adventitious roots. It is least prevalent in the leaves. Large tumours are formed, sometimes as much as 7 inches long by 4 inches broad, usually on the lower parts of the stem. To a lesser extent both male and female inflorescences may be partially replaced by irregular outgrowths full of black smut spores. The shape of the plant may indicate infection with smut, for infected flower heads become so heavy that they bend down in some unnatural position. When the stem is attacked deformations may arise both above and below the smutted area.

The smut tumours may assume various peculiar shapes. At first they are covered with a grey, or whitish, shining skin made up partly of plant tissue, partly of a membrane formed by the fungus. Until it ripens the blackish-brown spore mass inside is moist and greasy but it rapidly changes to a dry powder which spreads the disease far and wide. The infected plants and parts of plants should be removed and burnt as soon as they are seen and all debris of the crop should be destroyed in the autumn. Infection is spread by maize debris in manure.

It is stated abroad that the fungus is not carried by the seed, but its appearance in localities where maize has never been grown before indicates that this does occur to some extent. Possibly it is carried only on the surface of the seed. Attempts to breed resistant varieties are in progress in countries where maize is an important crop.

Chapter 2

Corn Disease Decreases Food Stock

WHEAT DISEASES

Many different types of diseases affect wheat. They can be classified as seed-borne diseases, leaf and head blight diseases, crown and root rot diseases, and virus diseases.

SEED-BORNE DISEASES

Most problems resulting from seed-borne diseases have been eliminated by highly effective seed-treatment fungicides. Several seed-borne diseases are of concern to wheat growers. They are seed-borne scab, seed-borne Stagonospora (previously known as Septoria), common bunt (stinking smut), and loose smut. Seed-borne scab and Stagonospora are discussed later in the section on leaf and head blight diseases. Both diseases result in lightweight, shriveled kernels that may be moldy.

Producers should follow the recommendations listed below for proper seed cleaning and planting to reduce seedling blight and stand losses resulting from planting diseased seed. The two smut diseases, stinking smut and loose smut, can be particularly devastating. Stinking smut causes losses by giving the seed of diseased plants a foul, fishy odor, making the grain unfit for milling.

Producers have been docked severely when attempting to sell smutty grain. Loose smut affects plants and yield by converting the grain and parts of the head to smut spores. Therefore, infected plants have no grain left to harvest. There are no varieties resistant to stinking smut, and numerous races of the loose smut fungus exist.

Note the following guidelines to control seed-borne diseases and seedling blights:

- Plant the highest-quality, disease-free seed possible. Seed-production fields should be inspected from the head-emergence growth stage through harvest for occurrence of scab, Stagonospora glume blotch, loose smut, and stinking smut. Do not use seed from severe smut infested fields.
- Clean seed thoroughly to remove all shriveled, lightweight kernels. This may require raising the test weight of the grain by several pounds per bushel.
- Have a standard germination test run on the seed. Use only seed with 80 per cent or greater germination percentage. If poor germination is due to Fusarium head scab, certain seed treatment fungicides can improve germination by 15 to 20 per cent.
- All wheat seed should be treated with a seed-treatment fungicide effective for control of smut fungi, Fusarium scab and Stagonospora.

Plant in a well-prepared seed bed or with a no-till drill capable of proper seed placement, when soil moisture is adequate and soil temperatures are not too high.

LEAF- AND HEAD- BLIGHT DISEASES

Major diseases in this group are powdery mildew, leaf rust, Septoria tritici leaf blotch, Stagonospora (Septoria) nodorum leaf and glume blotch, and Fusarium head scab. All can cause major yield losses, but their occurrence is essentially weather-dependent.

Cool, rainy weather from mid-April through the flowering period of the wheat plant in late May to early June favours the development of most of these diseases. They require either the leaf surfaces to be wet for a certain period of time or the relative humidity within the plant canopy to be near 100 per cent. Powdery mildew and Septoria tritici leaf blotch are the first leaf diseases to occur in the spring. Both are favoured by cool, humid, or wet weather. Leaf rust and Stagonospora nodorum leaf blotch require slightly warmer weather, thus they follow in mid to late May.

Stagonospora nodorum glume blotch and Fusarium head scab become evident in June soon after flowering, especially if wet weather persists through this time. The use of resistant varieties is the major control procedure for these disease management, leaf, and head blight diseases. Few varieties are resistant to all of these diseases. Determine which diseases cause most consistent problems in the area, and choose a variety based on its level of resistance.

The level of fungal carryover from one wheat crop to the next is minimal if a two-to three-year rotation away from wheat is maintained. Rotation is a primary control measure for powdery mildew and the Septoria and Stagonospora diseases. Spores of leaf rust are blown up from the southern states in late May; therefore, crop rotation has little effect on the incidence of leaf rust. Adequate and balanced fertility, to provide optimum nutrition for hardy plants, helps lessen the adverse effects of foliage diseases. High rates of nitrogen will favour powdery mildew and Stagonospora leaf and glume blotch.

Head scab is usually more severe when wheat is planted after corn because the fungus causing scab is the same one that causes Gibberella stalk rot. When possible, wheat should follow soybean or other legumes in the cropping sequence. Fungicides are available to control most foliar diseases of wheat. However, the use of these fungicides should be based on sound economic decisions and the level of disease in the field. Fungicides have been profitable during years when foliar diseases have been severe on susceptible cultivars or in locations where diseases are a persistent problem.

CROWN AND ROOT ROT DISEASES

Two of the more important crown- and root-rot diseases are take-all and Cephalosporium stripe. Both of these diseases are soilborne, meaning that the fungi that cause these diseases reside in the soil. Take-all can be recognized by examining the roots and lower stems of prematurely killed plants. The base of the stem and roots will have a scurfy, black appearance. Cephalosporium stripe disease is characterized by alternating yellow and brown stripes that extend the entire length of the leaf blades. Both of these diseases are favoured by planting wheat year after year in the same field. Usually a rotation sequence with a break of two years or more between wheat crops will eliminate fungal carryover.

The exception to this is when no-till is used to produce all crops in the rotation sequence and wheat residues do not decompose before the next wheat crop is planted. The other exception to this is when perennial grass weeds, such as quack grass, become established in the field. Proper rotations, tillage and elimination of grass weeds have been highly effective in managing both diseases. Additional control of the stripe disease can be achieved by maintaining a soil pH above 6.2 by proper liming according to a soil test. Adequate soil fertility will also reduce yield losses from root diseases.

VIRUS DISEASES

Wheat spindle streak mosaic (yellow mosaic) and barley yellow dwarf are the two most common virus diseases. Wheat spindle streak mosaic is a soilborne disease that is usually recognized in early May as the stems of the wheat plant begin to elongate. The upper leaves of affected plants will show short, spindle-shaped, yellow streaks. These symptoms may intensify if weather remains cool. The symptoms will tend to disappear as the weather begins to warm. Control of wheat spindle streak is achieved through the use of resistant varieties. A number of highly-resistant varieties are available. Barley yellow dwarf virus is transmitted by aphids.

Aphids arriving from the southern states transmit the virus to the newly-planted wheat crop in the fall. Severely affected plants may be stunted, have reddish or yellowish leaf tips and produce no heads. Yield losses greater than 50 per cent have occurred when entire fields have been infected in the fall. Because no varieties have an acceptable level of resistance and early-fall infections cause the greatest yield losses, wheat planting should be delayed until after the Hessian fly safe date when aphids have ended their fall flights.

CORN DISEASES

The major corn diseases can be grouped into four categories: leaf blights, stalk rots, ear rots, and viral diseases.

Leaf Blights: A number of leaf-blight diseases occur on corn. The most common are gray leaf spot, Stewart's bacterial leaf blight, and northern corn leaf blight. These diseases can be found in almost any field, depending on the year and susceptibility of the hybrid planted. Some leaf-blight diseases are most often found associated with continuous corn, especially in reduced-tillage, continuous corn fields.

These are anthracnose, gray leaf spot, eyespot, and northern leaf spot. All leaf blight diseases cause loss of green leaf tissue, resulting in fewer kernels and lightweight grain. Plants may be predisposed to stalk-rot diseases when leaf damage is severe. The amount of yield loss is usually related to the time when the plant's upper leaves become infected. The most severe yield loss occurs when the upper leaves, the ear leaf, and those above the ear, become infected at or soon after tasseling.

Yield losses will be minimal if disease does not occur on these leaves until six to eight weeks after tasseling. Leaf blight diseases are most effectively controlled by selecting hybrids with genetic resistance.

Contact seed dealer for information on hybrids with resistance to gray leaf spot, Stewart's bacterial leaf blight and other leaf diseases important in the area. A one-to two-year rotation away from corn and destruction of old corn residues by tillage may be helpful if susceptible hybrids must be grown. Fungicides are also available for control of leaf diseases, but are economically viable only under severe disease pressure.

STALK ROT

Stalk rots are the most important and common diseases of corn. Annual losses are estimated at 5 to 10 per cent. There are several stalk-rot diseases, but Gibberella stalk rot and anthracnose stalk rot currently are the most prevalent. Both are fungal diseases that result in premature ripening, chaffy ears, and lodging of plants before harvest.

The interior of the stalk becomes rotted, tissues break down, and the stalk is easily broken. Anthracnose stalk rot is usually associated with continuous corn and is recognized by the blackening of the outer surface of the stalk late in the season. Stalks with Gibberella stalk rot can be found in nearly any field.

Affected stalks often have pink to reddish discoloured internal tissues. Control of stalk rot diseases is based on reducing plant stress from factors such as lack of moisture, leaf diseases, insect injury, and nutritional stress. To reduce the affects of stalk rot diseases, follow as many of the following practices as possible:

- Select hybrids with good standability and resistance to leaf blight diseases.
- Adjust soil fertility to recommendations based on a soil test. Avoid excessive rates of nitrogen in relation to potassium.
- Follow a one- to three-year rotation away from corn. Soybeans forage legumes, and small grains are acceptable in the rotation. The longer the rotation away from corn the better.
- Plant at populations recommended for the hybrid grown. Overplanting leads to increased moisture, light and nutrient competition, and more plant stress.
- Harvest fields with the greatest level of rotted stalks first to avoid lost ears on lodged plants.
- Control insects, particularly root worms and stalk borer. Insects cause injuries to plant roots and stalks permitting stalk rot fungi to enter the plant.

EAR ROT

Gibberella, Fusarium, and Diplodia ear rot diseases, but Gibberella ear rot is the most important. The Gibberella ear rot fungus is the same fungus that causes Gibberella stalk-rot disease. Gibberella enters from the silk end of the ear when cool, wet weather persists for several weeks through late silking of the crop.

The occurrence of a whitish to pinkish mold on the ear tip is diagnostic, but extensive mold growth may not occur. On shelled grain, the symptoms may be seen as a pinkish colouration in some of the kernels. Even though extensive rotting does not always occur, the disease is serious because the fungus frequently produces toxins that makes the corn unfit for feeding.

Hogs are particularly sensitive to the toxins produced in moldy grain and may refuse to eat it even when hungry. If hogs refuse to eat grain, have a mycotoxin analysis run to determine the kinds and levels of toxin present. Some corn hybrids are less susceptible than others to Gibberella ear rot. Ears with tight husks which mature in an upright position often have more ear rot than those maturing in a declined position.

Diplodia ear rot appears to be more common in continuous corn under reduced tillage. Ears affected by Diplodia are covered with a thick mat of white fungal growth.

Fusarium ear rot is common, but only individual kernels are affected on ears. Plant hybrids known to be less susceptible to these ear diseases. Grain with evidence of ear or kernel rot should be dried to 14 per cent moisture before storage and maintained at this level until used. Feeding less than 5 per cent moldy kernels may prevent feeding problems, but grain should be tested for mycotoxin levels prior to feeding to avoid any problems.

VIRUS DISEASES

Maize dwarf mosaic and maize chlorotic dwarf are potentially destructive diseases where johnsongrass is established. The two viruses that cause these diseases are able to survive in this perennial weed grass. Aphids and leafhoppers feeding on johnsongrass in the spring pick up the virus and inoculate nearby corn. Control is achieved by planting resistant or tolerant hybrids. Efforts also should be made to eradicate johnsongrass.

SOYBEAN DISEASES

The major soybean diseases can be classified as root rots, stem rots, leaf blights, and seed diseases.

ROOT ROTS

Several root and lower-stem rot diseases are frequently encountered. These diseases are caused by fungi that live in the soil and parasitize the lower portion of the soybean plant. The two most common are Phytophthora root rot and Rhizoctonia root rot. Phytophthora root rot is the most destructive. Thousands of acres of soybeans are destroyed by this disease each year. The Phytophthora fungus attacks plants at any stage of growth, from germinating seedlings through mature plants.

Infected plants generally wilt and die soon after infection or are stunted. Phytophthora root rot occurs most frequently in heavy soils with poor drainage or in years with high rainfall. Fungal populations are comprised of a number of races that attack varieties with different genes for resistance. Rhizoctonia root rot is generally more prevalent in years when weather conditions are dry in early spring and then turn wet; however the disease can occur under other conditions, as well. Rhizoctonia attacks the base of the plant at the soil line causing reddish-brown cankers on the lower stem.

Charcoal rot has occurred in years with high temperatures during pod-filling. Temperatures at 85 to 95 degrees F for extended period's favours disease development as does drought stress.

Symptoms include stunting, yellowing of leaves, discoloured root tissues, and premature death. Diagnosis of charcoal rot is based on characteristic black zoning or charcoal-coloured speckling within root tissues. There are no adequate management methods for Rhizoctonia root rot and no resistant varieties are available for charcoal rot. Adequate fertility, a three-year crop rotation including small grains or corn may help in reducing losses.

Phytophthora root rot can be controlled with resistant soybean varieties. However, a number of different races occur, so previously resistant varieties may be susceptible to new races that develop within fields. Varieties with a high level of partial resistance are also available.

These varieties are susceptible to infection, but do not have the excessive yield loss associated with susceptible varieties. A listing of varieties with specific Rps resistance genes to the prevalent races of Phytophthora and ratings for the level of partial resistance to the disease. Performance of resistant and partial resistant varieties can be improved by treating seed with fungicides that control Phytophthora. Losses also can be reduced by using some form of tillage and providing adequate subsoil and surface drainage to problem fields.

STEM ROTS

The symptoms of most stem rots develop two to three weeks prior to maturity and often are misdiagnosed as early maturing beans. Cool, wet weather conditions in midsummer favour development of Sclerotinia white mold, brown stem rot, and Diaporthe stem rot which are all found. Hot, dry summer conditions promote the development of charcoal rot. Sclerotinia white mold is the most common of the stem rot diseases and has been increasing in incidence. This disease has a characteristic fluffy white mold growth that develops on the nodes and stems. The white mold continues to grow and causes a girdling lesion on the stem.

The leaves wilt and turn brown, affected plants die, but remain erect in the field. These dead plants with wilted leaves can be spotted in fields during mid- to late August and can be confused with brown stem rot. Black, hard sclerotia (resistant fungal bodies) develop both inside and outside the stems and within the pods. These sclerotia are harvested with the beans and can be collected with seed by combines, returned back to the soil with the debris, or carried from field to field on harvesting equipment.

Management of this disease is first dependent upon preventing introduction into a field. Plant only well-cleaned seed and treat seed with a fungicide effective against Sclerotinia. Some varieties do appear to be less susceptible to this pathogen by having less disease by the end of the season. It will be difficult to reduce the level of Sclerotinia in fields with high disease severities, but long crop rotations with corn, wheat, and/or alfalfa and selecting varieties that are less susceptible should improve the productivity of affected fields.

LEAF DISEASES

Several leaf diseases of soybean are common every year, but they seldom destroy enough leaf tissue to reduce yields. One of these diseases, Septoria brown spot, usually develops on the lowest leaves in late June and early July and results in a conspicuous yellowing followed by leaf-drop.

This disease is worse during wet weather and may reappear in late summer during wet, cooler weather and cause premature leaf yellowing and leaf-fall. Bacterial leaf blight occurs on leaves at various levels on the soybean plant, depending on the stage of growth in relation to the rain periods favouring disease establishment.

The blighted areas of the leaves frequently drop out or tear away giving a ragged appearance to the foliage. Downy mildew causes yellow spots on the upper sides of the leaves and grayish tufts of mold growth on the lower surfaces of leaves. Because very little or no yield loss results from the occurrence of these diseases, no control methods are recommended. Most varieties appear susceptible to these leaf diseases and continuous cropping to soybean may increase their severity.

SEED DISEASES

Moldy seed at harvest results from one or more damaging diseases of soybean caused by Phomopsis and Diaporthe fungi. Phomopsis seed decay is most serious when wet conditions prevail late in the season as the seeds mature and before they dry down to a moisture level suitable for combining. Soybeans left in the field because of delayed harvest due to wet weather are almost sure to develop Phomopsis infections. Moldy seed results from fungal invasion of pods and seeds.

The most important aspect of Phomopsis seed decay is

the effect it has on seed germination. Moldy seed will not germinate or seedlings die before emergence. Seeds that look healthy may often have seed coats colonized by Phomopsis. These fungi will grow into the germinating seed and kill the seedlings before emergence.

The level of infected seeds can be determined by a standard germination test. Producers interested in growing their own soybean seed should be particularly aware of the problems associated with Phomopsis seed decay. Producers should avoid growing continuous soybean and harvest those fields to be saved for seed first. Before planting, all seed should be evaluated using a standard germination test.

Use only those seed lots with 80 per cent or greater germination. On those lots with 80 to 90 per cent germination, adjust the seeding rate to compensate for the lower germination percentage, and plant only if adequate soil moisture is available for rapid germination. Most fungicide seed treatments available for use on soybean control Phomopsis. However, growers should not expect to increase germination of diseased seed lots by more than 20 per cent.

SOYBEAN CYST NEMATODE

Soybean cyst nematode (SCN) can be found at some level in most of the soybean producing counties of the state. Frequently no visible symptoms of nematode damage are observed other than reduced yield. Visible symptoms of SCN damage includes stunted, yellow plants and weed infestations where diseased soybeans lack adequate growth.

SCN is a microscopic round worm that exists in soil as eggs, worms, and cysts. Soils with moderate to high levels of nematodes (250 to 2,000 eggs/200 cc of soil) should implement control measures to reduce nematode populations and reduce yield losses. The only definite way to confirm the presence of SCN, or determine the level of infestation, is to submit soil samples for laboratory analysis.

ALFALFA DISEASES

There are more than a dozen diseases of alfalfa. They can

be grouped into two categories: those that affect the stems, crowns, or roots and those affecting the foliage. The stem and root-rot diseases are the most serious.

Major losses of stand have been caused by Phytophthora root rot, anthracnose, and Sclerotinia crown rot. Verticillium wilt is a constant threat because some farmers are still growing old varieties that lack resistance to this disease. Phytophthora root rot has been responsible for loss of stand in the seedling year when rainfall is above average and/or surface and subsurface drainage is poor.

It is caused by a soilborne fungus that becomes active under wet soil conditions and may attack both seedlings and older plants. Dark brown, decayed areas on the tap root two to three inches below the soil surface are early indicators of Phytophthora root rot. Soil moisture is the key factor affecting disease development. Any improvement in surface and subsurface drainage will reduce losses from this disease as well as from other less common root rots. Varieties with good levels of resistance are now available. Phytophthora specific fungicide seed treatments are available to help prevent loss of stands due to seedling damping-off.

Fungicides are labeled for use at planting either as a broadcast application or incorporated on fertilizer granules and applied in a band beneath the seed. Aphanomyces root rot may contribute to poor alfalfa establishment and reduced growth in wet soils. Seedlings may die (damping off) if infection occurs at an early stage of development. Older seedlings are yellowed and stunted.

When Aphanomyces and Phytophthora occur together, they form a destructive disease complex.. Alfalfa varieties with moderate to good levels of resistance are available for control of Aphanomyces root rot.

Anthracnose is a major cause of thinning of older alfalfa stands. Seedlings may be killed or crown and crown buds of older plants may be affected. Wilted and dead bleached stems are characteristic of the disease.

Diamond-shaped lesions (cankers) with light-brown centres and dark margins develop on the lower stems.

Anthracnose is a warm, wet-weather disease. Resistant varieties are available. Anthracnose has long been recognized as a destructive disease of red clover in the southern areas.

The use of resistant varieties of red clover controls this problem. Sclerotinia crown rot occurs almost exclusively on late-summer seedings, especially when minimum-tillage methods are used.

Affected plants wilt and the cottony-white mold changes to hard black bodies on the crowns or lower stems. This disease is usually seen during the cooler periods of the year. Crop rotation, with two to three years away from alfalfa, aids in control. Verticillium wilt may be spread from field to field with infested seed and in manure from animals fed infested hay. This soilborne disease usually does not become a problem until the third production year.

It can be recognized on scattered plants that become yellow and stunted, and then gradually die, leaving a thin, unproductive stand. Control is mainly through use of disease-free seed, resistant varieties, and crop rotation. There are several foliage diseases of alfalfa. Any of these may cause considerable loss of leaves during periods of prolonged wet or humid weather. Little can be done about these diseases in the year that they occur.

OAT DISEASES

In the past, oat diseases have resulted in considerable yield losses. Many of these diseases are no longer important since the development of resistant varieties. Diseases that have caused problems in recent years include loose smut, covered smut, and barley yellow dwarf virus.

Several new resistant varieties are available for control of barley yellow dwarf virus. Only seed guaranteed to be free of the smut diseases or seed treated with a fungicide to control smut diseases should be planted.

Chapter 3

Fruit Disease Causes Seasonal Loss

APPLE POWDERY MILDEW

Damage from powdery mildew attack results in stunted growth. The foliage becomes distorted and twig growth is reduced. Also, the fruit surface may become russetted or discoloured, and dwarfed. Heavily mildewed trees are weakened, and are more susceptible to other pests and winter injury.

SYMPTOMS

Powdery mildew may be found on buds, blossoms, leaves, twigs, and fruit. Symptoms first appear in the spring on the lower surface of leaves, usually at the ends of branches. Small, whitish felt-like patches of fungal growth appear and quickly cover the entire leaf. Diseased leaves become narrow, crinkled, stunted and brittle. By mid-summer, tiny, black round specks show up on the lower leaf surface, but more commonly on the twigs.

These are fungal fruiting bodies, but their importance in the disease cycle is probably minimal. The fungus spreads rapidly to twigs, which stop growing and become stunted. In some cases the twigs may be killed back. Leaves and blossoms from infected buds will be diseased when they open the next spring. Infected blossoms shrivel and produce no fruit. Fruit symptoms are not usually seen unless the disease has built up to high levels on susceptible cultivars. Diseased fruit has a fine network type surface blemish called russetting.

Causal Organism and Disease Cycle

Powdery mildew is caused by the fungus, Podosphaera leucotricha. Powdery mildew overwinters as fungal strands (mycelium) in vegetative or fruit buds which were infected the previous season.

Infected terminals may have a silvery gray colour, stunted growth, and a misshapen appearance and are more susceptible to winter kill than are noninfected terminals. Temperatures near -18 degrees F kill a majority of mildewed buds and the fungus within them.

Even at lower temperatures, however, some powdery mildew survives. As buds break dormancy, the powdery mildew fungus resumes growth and colonizes developing shoots causing primary infections.

The powdery white appearance on infected shoots consists of many thousands of spores which are responsible for spreading the fungus and causing secondary infections later in the growing season.

Secondary infections are important because they produce the overwintering infected buds. Secondary infections usually develop on leaves and buds before they harden off and may reduce the vigour of the tree.

Fruitlets may become infected shortly after bloom, resulting in a web-like russetting on the mature fruit. Powdery mildew infections occur when the relative humidity is greater than 90 per cent and the temperature is between 50-77 degrees F. The optimum temperature range for the fungus is 66-72 degrees F. Although high relative humidity is required for infection, the spores will not germinate if immersed in water.

Leaf wetting is, therefore, not conducive to powdery mildew development. This is quite different from most other foliar and fruit diseases caused by fungi that require free water for their spores to germinate and infect. Under optimum conditions, powdery mildew can be obvious to the naked eye 48 hr after infection.

About 5 days after infection, a new crop of spores is produced. Non-germinated powdery mildew spores can tolerate hot dry conditions and may persist until favourable conditions for germination occur. It is important to remember that powdery mildew can be a problem in drier growing seasons when other diseases are not a problem.

Control

- Apple varieties vary greatly in their susceptibility to powdery mildew. Jonathan, Granny Smith, Mutsu (Crispin), Rome, Cortland, Baldwin, Monroe and Idared are very susceptible and should be avoided if powdery mildew is a problem. Most other varieties may also be infected if inoculum is present and conditions are favourable for infection.
- Plant trees in sunny locations with good air drainage. This reduces the humidity around trees and reduces the chances of disease.
- Where powdery mildew is a problem, a good fungicide spray programme is generally required for control. This is especially true in commercial plantings.

FIRE BLIGHT OF APPLES, CRABAPPLES AND PEARS

Fire blight is a common and very destructive bacterial disease of apples and pears. The disease is so named because infected leaves on very susceptible trees will suddenly turn brown, appearing as though they had been scorched by fire. As a result of this disease, blight susceptible pear cultivars are no longer grown in many parts in the Midwest.

Damage and losses from fire blight on apple result from: death or severe damage to trees in the nursery; death of young trees in the orchard; delay of bearing in young trees due to frequent blighting of shoots and limbs; loss of limbs or entire trees in older plantings as the result of girdling by fire blight cankers; and direct loss of fruit due to blighting of blossoms and young fruit.

Fire blight may cause severe damage to many other members of the Rosaceae family. Quince, crabapple, mountain ash, spirea, hawthorn, pyracantha, and cotoneaster are all susceptible. Cultivars within some of these species are resistant.

SYMPTOMS

Blossom and twig blight symptoms appear in the spring. Diseased blossoms become water-soaked and turn brown. The bacteria may then grow down into the blossom bearing twigs (spurs). Leaves on the spur become blighted, turning brown on apple and black on pear. Droplets of milky tan-coloured bacterial ooze may be visible on the surface of diseased tissue. These droplets contain millions of bacteria which can cause new infections. Twig blight starts at the growing tips of shoots and moves down into older portions of the twig. Blighted twigs first appear water-soaked, and then turn dark brown or black. Blighted leaves remain attached to the dead branches through the summer.

The end of the branch may bend over, resembling a shepherd's crook or an upside down "J". As the fire blight bacteria move through blighted twigs into main branches, the bark sometimes cracks along the margin of the infected area on the main branch causing a distinct canker. Both apple and pear fruit may be blighted. Rotted areas turn brown to black and become covered with droplets of ooze. The fruit remains firm but later dries out and shrivels into mummies.

Causal Organism

Fire blight is caused by the bacterium, Erwinia amylovora. The fire blight bacteria overwinter in living tissue at the margins of cankers on the trunk and main branches. The bacteria become active in the spring when temperatures get above 65 degrees F. Their growth is favoured by rain, heavy dews, and high humidity. By the time trees are blossoming, droplets of ooze containing the bacteria are present on the surface of cankers. Relatively few overwintering cankers become active and produce bacteria in the spring, but a single active canker may produce millions of bacteria, enough to infect an entire orchard.

The bacteria in droplets of ooze are spread by splashing rain or insects (mostly bees, flies, and ants) to open blossoms. The bacteria multiply rapidly in the blossom nectar, and invade the blossom tissue through natural openings called nectaries. The optimum temperature range for blossom blight infection is 65 to 86 degrees F. The bacteria are spread from blossom to blossom by rain or pollinating insects. Actively growing shoot tips are infected by bacteria that have been spread by rain or insects from both cankers and infected blossoms.

Invasion can occur directly through natural openings, such as lenticels and stomata, under conditions of prolonged rain and high humidity. However, shoot infection more commonly occurs through wounds created by sucking insects, such as aphids, leafhoppers, and tarnished plant bugs; by wind whipping; or by hail.

Fire blight bacteria multiply rapidly within an infected shoot. Droplets of ooze can form on the shoots within 3 days. Shoots remain highly susceptible to infection until vegetative growth ceases and the terminal bud is formed.

Control

Fireblight is one of the most difficult diseases of apple to control, and there is no one procedure that will give complete control. Though control is not an easy task, the use of several practices in an integrated manner should result in minimal damage from fire blight:

- Plant apple, crabapple, and pear varieties that are less susceptible to fire blight. Table gives a listing of the relative susceptibility of some of the more common apple and pear varieties. Fireblight is not as severe a disease problem on most crabapple varieties. A few crabapple varieties which can develop severe fireblight include: Silver Moon, Snowdrift, Red Jade, and Van Esseltine.
- Prune out fire blight cankers and blighted twigs. To decrease the inoculums level for the following season, prune out blighted twigs and cankers during the dormant season. During the dormant season (winter) there is much less chance of spreading bacteria. Branches that are more than half-girdled by cankers should be removed.

Cut off blighted twigs by making cuts at least 4 inches below the visible dead wood. Cankers can be cut out of trunks or large branches by removing dead tissue down to wood that appears healthy. If blighted twigs are pruned out during summer, cuts should be made 12 to 15 inches below diseased wood and pruning tools should be disinfested by dipping in a 2:10 solution of household bleach in water after each cut. We recommend that commercial growers do a thorough job of pruning out blighted wood in the dormant season and not in summer.

- Follow proper pruning and fertilization practices. Excessive nitrogen fertilizer and heavy pruning will promote vigorous growth of succulent tissue which is more susceptible to fire blight. Adjust management practices on susceptible varieties to promote moderate growth. Make fertilizer applications in early spring or late fall after growth has ceased.
- Sucking insects create wounds through which fire blight bacteria can enter. These pests should be controlled throughout the growing season. To protect bees, do not apply insecticides during bloom.

SCAB OF APPLE AND CRABAPPLE

Apple scab is one of the most serious diseases of apple and ornamental crabapple. Disease development is favoured by wet, cool weather that generally occurs in spring and early summer.

Both leaves and fruit can be affected. Infected leaves may drop resulting in unsightly trees, with poor fruit production. This early defoliation may weaken trees and make them more susceptible to winter injury or other pests. Infected fruits are blemished and often severely deformed. Infected fruits may also drop early.

Symptoms

Symptoms first appear in the spring as spots (lesions) on the lower leaf surface, the side first exposed to fungal spores as buds open. At first, the lesions are usually small, velvety, olive green in colour, and have unclear margins. On some crabapples, infections may be reddish in colour. As they age, the infections become darker and more distinct in outline. Lesions may appear more numerous closer to the mid-vein of the leaf. If heavily infected, the leaf becomes distorted and drops early in the summer.

Trees of highly susceptible varieties may be severely defoliated by mid to late summer. Fruit symptoms are similar to those found on leaves. The margins of the spots, however, are more distinct on the fruit. The lesions darken with age and become black and "scabby." Scabs are unsightly, but are only skin deep. Badly scabbed fruit becomes deformed and may fall before reaching good size.

Causal Organism

Apple scab is caused by the fungus, Venturia inaequalis. It survives the winter in the previous year's diseased leaves that have fallen under the tree. In the spring, the fungus in old diseased leaves produces millions of spores.These spores are released into the air during rain periods in April, May and June. They are then carried by the wind to young leaves, flower parts and fruits. Once in contact with susceptible tissue, the spore germinates in a film of water and the fungus penetrates into the plant.

Depending upon weather conditions, symptoms (lesions) will show up in 9 to 17 days. The fungus produces a different kind of spore in these newly developed lesions. These spores are carried and spread by splashing rain to other leaves and fruits where new infections occur.

The disease may continue to develop and spread throughout the summer. Because a film of water on leaves and fruit is required for infection to occur, apple scab is most severe during years with frequent spring rains.

Control

- The use of resistant or scab immune varieties is the ideal method for controlling scab. Currently there are several apple varieties that are totally resistant to scab. Backyard growers are strongly encouraged to consider using these resistant varieties in order to reduce or eliminate the need for fungicide applications around the home. Scab resistant apple varieties include: Prima, Priscilla, Sir Prize, Freedom, Liberty, Jonafree, Enterprise, Goldrush, Redfree, Pristine, Williams Pride, Novamac and Nova Easygro.

All other varieties, including most commercially grown varieties are susceptible to scab; however, they differ in their degree of susceptibility. McIntosh, Cortland, Red Delicious and Rome Beauty are all very susceptible to scab. Golden Delicious and Jonathan are less susceptible. Lists of scab resistant ornamental crabapples are available from many nurseries and garden centres.

- Rake and destroy fallen leaves. This will reduce the number of spores that can start the disease cycle the next year. Where resistance to scab is not present, fungicide application is the primary method of control.

GRAPE BLACK ROT

Black rot is one of the most damaging grape diseases. All cultivated varieties of grapes are susceptible to infection by the black rot fungus. If not controlled, some or all of the grapes within a cluster will be rotted.

The disease is favoured by warm, humid weather as is found during the summer. Before good control measures were devised, vineyards along the River often were hard hit. Grape growers commonly lost most of their crop, and the grape industry was literally driven out of the area.

SYMPTOMS

Symptoms of black rot first appear as small yellowish spots on leaves. As the spots (lesions) enlarge, a dark border forms around the margins. The centres of the lesions become reddish brown. By the time the lesions reach 1/8 to 1/4 inch in diameter (approximately two weeks after infection), minute black dots appear. These are fungal fruiting bodies (pycnidia) and contain thousands of summer spores (conidia). Pycnidia are often arranged in a ring pattern, just inside the margin of the lesions. Lesions may also appear on young shoots, cluster stems, and tendrils.

The lesions are purple to black, oval in outline, and sunken. Pycnidia also form in these lesions. Fruit symptoms often do not appear until the berries are about half grown. Small, round, light-brownish spots form on the fruit. The rotted tissue in the spot softens, and becomes sunken. The spot enlarges quickly, rotting the entire berry in a few days. The diseased fruit shrivels, becoming small, hard, black and wrinkled (mummies). Tiny black pycnidia are also formed on the fruit mummies. The mummies usually remain attached to the cluster.

Causal Organism

Grape black rot is caused by the fungus, Guignardia bidwellii. Black rot survives the winter in cane and tendril lesions and fruit mummies. In the spring during wet weather, the pycnidia on infected tissues absorb water and conidia are squeezed out.

Conidia are splashed about randomly by rain and can infect any young tissue in less than 12 hours at temperatures between 60-90 degrees F.

A film of water on the vine surface is necessary for infection. A second type of spore, an ascospore, may also be produced in overwintered fruit mummies. Ascospores are forcibly discharged into the air and can travel considerable distances. Research has shown that ascospores are an important source of primary infections in the spring.

Control

- Sanitation is important. Destroy mummies, remove diseased tendrils from the wires, and select fruiting canes without lesions. It is very important not to leave mummies attached to the vine. Research has shown that mummies on the ground release most or all of their ascospores before the end of bloom. Mummies left up in the trellis can produce ascospores and conidia throughout the growing season, thus making control of this disease much more difficult. If only a few leaf lesions appear in the spring, remove these infected leaves.

- Plant grapes in sunny open areas that allow good air movement. Proper row orientation to prevailing winds and good weed control beneath the vines also enable plants to dry more quickly during wet weather.

PEACH CANKER

Peach canker is a fungus disease common on apricot, prune, plum, and sweet cherry trees as well as on peach trees. The disease is common in peach orchards and is a frequent cause of limb dying and death of peach trees. Other common names for peach canker are perennial canker and Valsa canker. The fungi that cause this disease enter the plant through wounds. Infection results in dead and weakened twigs. and branches, and in reduced yields.

SYMPTOMS

The first symptoms appear in early spring as gummy drops of sap around wounded bark.

The diseased inner bark begins to break down, causing the cankered surface to appear depressed. Black specks, which are fungal spore producing bodies, appear on the bark surface or under the bark tissue. During wet periods spores ooze out of these fungal bodies in tiny orange or amber coloured, curled strands. During the summer, healthy bark (callus tissue) grows over the edges of the narrow, oval shaped cankers. In the fall, the fungus resumes growth, attacking and killing the new callus tissue. Over a period of years, a series of dead callus ridges form as the canker gets larger. Eventually, the canker may completely surround a branch.

The portion of the branch beyond the canker then dies. Large amounts of gum are usually produced around cankered areas. Peach canker is often confused with other problems which cause cankering and gumming. Among these are bacterial canker, insect borer injury and mechanical injury. When insects are involved, chewed-up wood dust is usually visible under the gum. Mechanical injury can often be verified by carefully reviewing recent operations in the area.

Causal Organism

Peach canker is caused by the fungi, Cytospora leucostoma and Cytospora cincta.

These fungi are weak pathogens and generally do not attack healthy, vigorous peaches bark. Winter injury, insect damage, and mechanical injury are common types of wounds serving as entry points. The fungi survive the winter in cankers or in dead wood. During spring and summer, spores produced in the cankers are spread by wind and rain to wounds on the same or nearby trees. The spores are not blown over long distances in the wind.

Infection and canker development depend on temperature and the species of fungus involved. Cytospora cincta is favoured by lower temperatures than Cytospora leucostoma. Because of the manner of infection and development of this disease, no single control measure is adequate. Most known control methods act indirectly by reducing points of entry or by reducing the level of inoculum. Fungicides are generally ineffective for controlling this disease.

Control

- Prune young trees carefully to avoid weak, narrow-angled crotches. Narrow-angled crotches are frequent sites of breakage and winter injury.
- Delay pruning until early spring. This promotes quick healing. Remove cankered branches and dead wood while pruning. Do not leave protruding pruning stubs. Cut flush to the next larger branch.
- Eradicate cankers and remove badly cankered limbs,

branches or trees. Burn or remove all cankered limbs soon after pruning. These limbs or branches serve as a reservoir for the disease causing fungi. Sanitation is critical, especially during the early life of the orchard.
- Do not plant new peach trees near established trees with canker.
- Avoid mechanical and insect injury.
- Promote vigorous, healthy peach trees with proper fertilization, pruning, and water.
- Do not over-fertilize late in the season. Winter injury is more common on these trees because winter hardening is delayed.
- White latex paint applied to the southwest side of trunks and lower scaffold branches may help avoid cold injury.
- Maintain a good control programme for other diseases and insect pests, especially borers.

PEACH LEAF CURL

Leaf curl is a springtime disease that occurs on peach, nectarine and related ornamental plants. The disease, though not a problem every spring, can be severe during cool, wet springs that follow mild winters. The leaf curl fungus damages peach trees by causing an early leaf drop. This weakens the trees, making them more susceptible to other diseases and to winter injury. Weakened trees also will produce fewer fruit the following season. Yield may be further reduced when blossoms and young fruit become diseased and drop.

SYMPTOMS

Symptoms of leaf curl appear in the spring. Developing leaves become severely distorted, and have a reddish or purple cast. Later, as spores form on the leaf surface, the leaves become powdery gray in colour. Shortly after this, the leaves turn yellow or brown and drop. There is no secondary spread of this disease from leaves infected in the spring to new leaves produced later in the growing season. Once infected leaves drop, no further symptoms will appear during that growing season. Diseased twigs become swollen and stunted, and may have a slight golden cast. They usually produce curled leaves at their tips. Though rarely seen, flowers and fruit may also become diseased. They drop shortly after they are infected. Diseased fruit has shiny, reddish, raised, warty spots.

Causal Organism

Peach leaf curl is caused by the fungus, Taphrina deformans. The fungus survives the winter as spores (conidia) on bark and buds. Infection occurs very early in the growing season. During cool, wet spring weather the conidia infect new leaves as they emerge from the buds. Host plant tissues are susceptible for only a short period. As the tissues mature they become resistant. The fungus produces another type of spore (ascospore) on the upper surface of the diseased leaves. During wet weather, ascospores produce additional conidia by budding. These conidia are carried to other parts of the tree by rain and wind, where they will overwinter until the next spring. Environment can limit leaf curl infection.

This partially explains why the disease does not occur every year. Leaf curl is worse when the weather is cool and wet. Low temperatures are thought to retard maturation of leaf tissue, thus prolonging the time infection may occur. The fungus can penetrate young peach leaves readily at temperatures between 50 and 70 degrees F, but only weakly below 45 degrees F. Rain is necessary for infection.

Control

Leaf curl is not difficult to control. Since the fungus survives the winter on the surface of twigs and buds, a single fungicide spray, thoroughly covering the entire tree, will provide control. If leaf curl does result in significant defoliation in the spring, the fruit on affected trees should be thinned to compensate for the loss of leaves. Over-cropping the tree will weaken it and make it more susceptible to winter injury.

ANTHRACNOSE OF RASPBERRY AND BLACKBERRY

Anthracnose is a disease common to raspberries,

blackberries, and other brambles or cane fruits. It causes severe damage to black and purple raspberries and susceptible varieties of red raspberries throughout the United States. The disease reduces the size and quality of fruit on infected canes. In addition, it may kill canes or weaken them so that they do not survive the winter. Other common names for this disease are "cane spot" and "gray bark."

SYMPTOMS

Anthracnose first appears in the spring on the young shoots as small, purplish, slightly raised or sunken spots. Later, they enlarge and become ash gray in the centre with slightly raised purple margins. The spots are often so close together on black and purple raspberries that they form large irregular areas (cankers). The cankers may encircle the cane, sometimes causing the death of the cane beyond the canker. The bark in badly cankered areas often splits. Late season infections result in superficial gray, oval spots. The spots have definite margins, but are not sunken.

They may become so numerous that the spots blend together, covering large portions of the cane. This is the characteristic "gray bark" symptom which is common on red raspberry. Dark coloured specks (fungal fruiting bodies) develop in circles on the gray bark. Anthracnose sometimes attacks the leaves and can cause some leaf drop. Small spots, about 1/16 inch in diameter, with light gray centres and purple margins appear on the leaves. Lesion centres later fall out, leaving a shot hole effect.

Causal Organism

Anthracnose is caused by the fungus Elsinoe veneta. The fungus survives the winter in lesions on diseased canes. The following spring and summer, during wet and rainy periods, spores are released.

Spores are carried by splashing rain to healthy first-year primocanes. These spores may then germinate and infect young tissues on developing primocanes. Disease development is favoured by extended periods of wet weather.

- All steps possible should be taken to improve air circulation within a planting, to allow faster drying of foliage and canes. Reducing the number and duration of wet periods should reduce the potential for infection. Excessive applications of fertilizer (especially nitrogen) should be avoided, since it promotes excessive growth of very susceptible succulent plant tissue. Plants should be maintained in narrow rows and thinned to improve air circulation and allow better light penetration. Weeds are very effective in reducing air movement; therefore, good weed control within and between rows is important for improving air circulation within the planting.
- After harvest, remove and destroy all old fruited canes (floricanes) and any new primocanes that are infected. It is best to remove old canes shortly after harvest, and it is critical to have them removed before new growth starts in the spring. Remember, the fungus overwinters on old-infected canes.
- Remove all wild brambles growing in the area because they can serve as a reservoir for the disease.
- Where the disease is established in the planting, fungicide applications are generally required to achieve adequate control.

SPUR BLIGHT OF RED RASPBERRIES

Spur blight is caused by the fungus Didymella applanata. Spur blight occurs only on red and purple raspberries. Spur blight has been considered to be a serious disease of red raspberry; however, recent studies in Scotland suggest that spur blight actually does little damage to the cane. The extent of damage caused by spur blight in the United States is not clearly understood.

The spur blight fungus has been reported to reduce yields in several ways. It can blight the fruit bearing spurs that are produced on the side branches, cause premature leaf drop, and kill buds on the canes that later develop into fruit bearing side branches. In addition, berries produced on diseased canes may be dry, small and seedy.

SYMPTOMS

The symptoms first appear on young first-year primocanes in late spring or early summer. Purple to brown areas (lesions) appear just below the leaf or bud, usually on the lower portion of the stem. These lesions expand, sometimes covering all the area between two leaves. In late summer or early fall, bark in the affected area splits lengthwise and small black specks, which are fungal fruiting bodies (pycnidia) appear in the lesions.

They are followed shortly by many slightly larger, black, erupting spots; another form of fungal fruiting body (perithecia).

Leaflets sometimes become infected and show brown, wedge-shaped diseased areas, with the widest portion of the wedge toward the tip of the leaf. Infected leaflets may fall off, leaving only petioles without leaf blades attached to the cane. When diseased canes become fruiting floricanes during the next season, the side branches growing from diseased buds are often weak and withered.

Causal Organism

Spur blight is caused by the fungus, Didymella applanata. It survives the winter in lesions on diseased canes. The following spring and summer, during wet and rainy periods, spores are released and carried by splashing rain and wind to nearby primocanes. There they germinate in the presence of water and produce new infections, where the fungus will again over winter.

Control

All steps possible should be taken to improve air circulation within a planting, to allow faster drying of foliage and canes. Reducing the number and duration of wet periods should reduce the potential for infection. Excessive applications of fertilizer (especially nitrogen) should be avoided, since it promotes excessive growth of very susceptible succulent plant tissue.

Plants should be maintained in narrow rows and thinned to improve air circulation and allow better light penetration. Weeds are very effective in reducing air movement; therefore, good weed control within and between rows is important for improving air circulation within the planting. Wild brambles, especially wild red raspberries, growing in the area should be removed. They can serve as a reservoir for the disease.

After harvest, remove and destroy all old fruited floricanes and any first-year primocanes that are infected. It is important that old canes be removed before new canes emerge in the spring. If spur blight becomes an important problem in the planting, growers may want to consider the use of fungicide.

Special fungicide sprays specifically for control of spur blight are generally not warranted.

BROWN ROT OF STONE FRUITS

Brown rot is a common and destructive disease of peach and other stone fruits (plum, nectarine, apricot, and cherry). The brown rot fungus may attack blossoms, fruit, spurs (flower and fruit bearing twigs), and small branches. The disease is most important on fruits just before ripening, during and after harvest. Under favourable conditions for disease development, the entire crop can be completely rotted on the tree. Peaches not kept in cool storage may be rotted in two to three days by the fungus.

SYMPTOMS

The symptoms of brown rot are very similar on all stone fruit. Symptoms first appear in the spring as the blossoms open. Diseased flowers wilt, turn brown, and may become covered with masses of brownish-gray spores. The diseased flowers usually remain attached into the summer. Young fruits are normally resistant, but may become infected through wounds. As fruits mature they become more susceptible to attack, even in the absence of wounds. Fruit infections appear as soft brown spots which rapidly expand and produce a tan powdery mass of conidia. The entire fruit rots rapidly, then dries and shrinks into a wrinkled "mummy." Rotted fruit and mummies may remain on the tree or fall to the ground. Fruit infection may spread rapidly, especially if environmental conditions are favourable and fruits are touching one another.

The fungus may move from diseased blossoms or fruit into the spurs. The fungus may then invade and cause diseased areas (cankers) on the twigs below. Succulent shoots are sometimes infected by direct penetration near their tip. A canker may form encircling the twig, causing death of the twig beyond the canker (twig blight).

Causal Organism

Brown rot is caused by the fungus, Monilinia fructicola. The brown rot fungus survives the winter in mummified fruits (either on the ground or still on the tree) and in twig and branch cankers produced the preceding year. Both sources may produce spores that can infect blossoms and young shoots. At about blossom time, a mummified fruit that has fallen on the ground produces up to 20 or more small, tan, cup-like structures on slender stalks that are called apothecia.

As an apothecium matures, it becomes thicker and the cup opens to a bowl-like disc 1/8 to 1/2 inch in diameter across the top. The inner surface of each bowl is lined with thousands of spore-containing sacs (asci). At this stage, the slightest disturbance of air movement will cause an apothecium to forcibly discharge millions of spores.

These spores (ascospores) are carried by wind to the open or unopened blossoms and young shoots. If a film of water (either from dew or rain) is present for 5 hours or longer, the spores can germinate and penetrate the plant. Infected blossoms soon wilt and tan-gray tufts, composed of masses of another type of spore (conidia), develop on the outside of the flower shuck.

If the infected blossom does not drop off, the fungus soon grows through the pedicel to the twig and forms a canker. Masses of conidia are soon produced on the newly cankered twig surface during moist periods throughout May and June. These summer spores are easily detached, and, like the ascospores, are mainly wind-borne. They are also splashed by rain or carried by insects to the growing fruit. Brown rot conidia can germinate and infect at temperatures of 32 to 90 degrees F. Wet weather and temperatures ranging from 60 to 70 degrees F are most favourable for disease development.

Following spring and summer rainy periods, mummified fruit still hanging in the tree become covered with masses of conidia that may result in blossom blight or fruit rot.

Mummies hanging in the tree do not produce ascospores. Although the flesh of young fruit is very susceptible to brown rot infection, the fruit has such a tough skin that the germ tubes of the summer spores do not normally penetrate. For this reason, young uninjured fruits are fairly safe from infection. However, any type of injury to the fruit will provide entry points for brown rot spores. Insect and hail wounds, fruit cracking, limb rubs, twig punctures, and a variety of picking and packing injuries greatly increase the losses due to brown rot.

Growers must realise that brown rot spores are practically everywhere during the fruit-ripening period. Infection is almost certain to occur if the weather is moist and if the fruit skin is broken in some way.

Control

- Sanitation is very important in controlling brown rot. All dropped and rotted fruit should be picked up and destroyed promptly. At the same time, remove all mummies from the trees. Prune out all cankers during the dormant season. Overripe or rotting fruit in the packing shed should be removed and destroyed at once.
- Control of insects that feed on fruit is essential. Remember that anything that causes wounding of the fruit will increase the incidence of brown rot. Special care should be taken during harvesting and packing to prevent puncturing or bruising of ripe fruit.

- Remove wild or neglected stone fruit trees that serve as reservoirs for the disease.
- Fruit should be cooled and refrigerated (as close to 32 degrees F as possible) immediately after harvest.
- The use of fungicide is an important part of the disease management programme for brown rot.

ORANGE RUST OF BRAMBLES

Orange rust is the most important of several rust diseases that attack brambles. All varieties of black and purple raspberries and most varieties of erect blackberries and trailing blackberries are very susceptible. Orange rust does not infect red raspberries.

Unlike all other fungi that infect brambles, the orange rust fungus grows "systemically" throughout the roots, crown and shoots of an infected plant, and is perennial inside the below ground plant parts. Once a plant is infected by orange rust, it is infected for life. Orange rust does not normally kill plants, but causes them to be so stunted and weakened that they produce little or no fruit.

SYMPTOMS

Orange rust-infected plants can be easily identified shortly after new growth appears in the spring. Newly formed shoots are weak and spindly.

The new leaves on such canes are stunted or misshapen and pale green to yellowish. This is important to remember when one considers control, because infected plants can be easily identified and removed at this time. Within a few weeks, the lower surfaces of infected leaves are covered with blister-like pustules that are waxy at first but soon turn powdery and bright orange.

This bright orange, rusty appearance is what gives the disease its name. Rusted leaves wither and drop in late spring or early summer. Later in the season, the tips or infected young canes appear to have outgrown the fungus and may appear normal. At this point, infected plants are often difficult to identify. In reality, the plants are systemically infected, and in the following years, infected canes will be bushy and spindly, and will bear little or no fruit.

Causal Organism and Disease Development

Orange rust is caused by two fungi that are almost identical, except for a few differences in their life cycles.

Arthuriomyces peckianus occurs primarily in the northeastern quarter of the United States and is the causal agent for the disease. Gymnoconia nitens is a microcyclic (lacks certain spores)

Control

- Whenever possible, start with disease-free, certified nursery stock.
- When diseased plants first appear in early spring, dig them out (including roots) and destroy them before pustules form, break open, and discharge the orange masses of spores. If plants are not removed, these spores will spread the disease to healthy plants.
- Remove all wild brambles from within and around the planting site. Wild brambles serve as a reservoir for the disease.
- Maintain good air circulation in the planting by pruning out and destroying old fruited canes immediately after harvest, thinning out healthy canes within the row, and keeping the planting free of weeds.
- Fungicide sprays are generally not considered an effective control method for orange rust.

BLACK KNOT OF PLUMS AND CHERRIES

Black knot of plums and cherries is a widespread and serious disease throughout the United States. Black knot is a common disease on wild plums and cherries and in home orchards where pruning and spraying are not regularly practiced.

The disease becomes progressively worse during each growing season and unless effective control measures are taken, it can stunt or kill the tree. The black knot fungus can infect American, European, and Japanese varieties of cultivated plums and prunes. Sweet, tart, and Mahaleb cherries are also affected by the fungus, but are generally less susceptible than plum or prune. Occasionally, it may also infect apricots, peaches and other Prunus species.

SYMPTOMS

The black knot fungus mainly affects twigs, branches, and fruit spurs. Occasionally, trunks may also become diseased. Usually, infections originate on the youngest growth. On infected plant parts, abnormal growth of bark and wood tissues produce small, light-brown swellings that eventually rupture as they enlarge. In late spring, the rapidly growing young knots have a soft (pulpy) texture and become covered with a velvety, olive-green growth of the fungus. In summer, the young knots turn darker and elongate.

By fall, they become hard, brittle, rough and black. During the following growing season, the knots enlarge and gradually encircle the twig or branch. The cylindrical or spindle-shaped knots may vary from one-half inch to a foot or more in length and up to 2 inches in diameter.

Small knots may emerge from larger knots forming extensive galls. After the second year, the black knot fungus usually dies and the gall is invaded by secondary fungi that give old knots a white or pinkish colour during the summer. Smaller twigs usually die within a year after being infected. Larger branches may live for several years before being girdled and killed by the fungus.

The entire tree may gradually weaken and die if the severity of the disease increases and effective control measures are not taken.

Causal Organism and Disease Development

Black knot is caused by the fungus, Dibotryon morbosum. The fungus overwinters in knots on twigs and branches or in the infected wood immediately surrounding them. In the spring, the fungus produces spores (ascospores) in sacs contained within tiny fruiting bodies on the surface of the knots. These ascospores are ejected into the air during rainy periods and are blown for moderate distances by wind currents. Only succulent green twigs of the current season's growth are susceptible to infection.

Ascospores that land on them may germinate and cause infection if the twigs remain wet for a sufficient length of time. Normal growth is disrupted in the infected regions, and a knot is formed as the fungus causes the plant to produce tumorlike growths. Knots may become visible by the late summer of the year of infection but often are not noticed until the following spring, when they begin to enlarge rapidly. New ascospores capable of spreading the disease may be formed in the young knots the year following infection but often are not formed until the second spring.

The fungus continues to grow in infected wood during the spring and fall months, causing the knots to elongate several inches each year and eventually girdle affected twigs and branches. Ascospores are potentially available from the time of bud break until terminal shoot growth stops, but the greatest number appear to be released during the period between white bud and shuck split. Although the precise environmental conditions required for infection are uncertain, only a few hours of rain apparently are required at temperatures above 55 degrees F (13 degrees C), whereas much longer rainy periods are required to produce infection at temperatures below this threshold.

Control

Most plum varieties grown, including Stanley and Damson, are susceptible to this disease. It has been reported that Early Italian, Brodshaw, Fallenburg, Methley and Milton are somewhat less susceptible than Stanley; Shiro, Santa Rose, and Formosa are much less susceptible; and President is apparently resistant to black knot. Japanese varieties of plums are generally less susceptible than most American varieties. When planting new plum or prune trees, avoid planting trees next to or downwind from an old or abandoned orchard with a significant black knot problem. Similarly, remove all wild plum and cherry trees (potential disease reservoirs) from fence rows or woodlands next to the orchard site.

Established orchards or backyard trees should be scouted or examined each year for the presence of black knot, and infected twigs should be pruned out and destroyed or removed before bud break.

Chopping prunings with a flail mower (to strip infected bark and knots from wood) is an alternative method of destroying infected prunings if burning or burying is impractical. It is important to prune at least 2-4 inches (5-10 cm) below each knot because the fungus grows beyond the edge of the knot itself.

If pruning is not possible because knots are present on major scaffold limbs or the trunk, they can be removed by cutting away the diseased tissue down to healthy wood and out at least 1/2 inch (1 cm) beyond the edge of the knot. Fungicides can offer significant protection against black knot, but are unlikely to be effective if pruning and sanitation are ignored. The timing of fungicide sprays should be adjusted to account for inoculum levels and weather conditions. Where inoculum is high because of an established black knot problem or a neighbouring abandoned orchard, protection may be needed from bud break until early summer.

Where inoculum has been maintained at low to moderate levels, sprays are most likely to be useful from white bud through shuck split (the period of maximum availability of ascospores). Fungicides are most necessary and will provide the greatest benefit if applied before rainy periods, particularly when temperatures are greater than 55 degrees F (13 degrees C). In evaluating control programmes, remember that knots often do not become apparent until the year following infection.

VERTICILLIUM WILT OF STRAWBERRY

Verticillium wilt of strawberry can be a major factor limiting production. When a plant is severely infected by the Verticillium wilt fungus, the probability of it surviving to produce a crop is greatly reduced. The Verticillium fungus can infect about 300 different host plants, including many fruits, vegetables, trees, shrubs and flowers, as well as numerous weeds and some field crops. The fungus can survive in soil, and, once it becomes established in a field or garden, it may remain alive for 25 years or longer. Cool, overcast weather interspersed with warm, bright days is most favourable for development of Verticillium wilt. Infection and disease development may occur when soil temperature is from 70 to 75 degrees F (21 to 24 degrees C).

Many soils contain the Verticillium wilt fungus. The fungus can be introduced into uninfested soil on seed, tools and farm machinery, and in the soil and roots of transplants.

SYMPTOMS

The first symptoms of Verticillium wilt in new strawberry plantings often appear about the time runners begin to form. In older plantings, symptoms usually appear just before picking time. Symptoms on above-ground plant parts may differ with the susceptibility of the cultivar affected. In addition, above-ground symptoms are difficult to differentiate from those caused by other root infecting fungi. Isolation from diseased tissue and culturing the fungus in the laboratory are necessary for positive disease identification. On infected strawberry plants, the outer and older leaves droop, wilt, turn dry and become reddish-yellow or dark brown at the margins and between veins.

Few new leaves develop, and those that do tend to be stunted and may wilt and curl up along the midvein. Severely infected plants may appear stunted and flattened, with small yellowish leaves. Brownish to blueish-black streaks or blotches may appear on the runners or petioles. New roots that grow from the crown are often dwarfed with blackened tips. Brownish streaks may occur within the decaying crown and roots. If the disease is serious, large numbers of plants may wilt and die rapidly. When the disease is not so serious, an occasional plant or several plants scattered over the entire planting may wilt and die.

Casual Organism

Verticillium wilt is caused by the soilborne fungus fumigant that is most effective is a mixture of chloropicrin (tear gas) 33 per cent and methyl bromide 66 per cent.

DOWNY MILDEW OF GRAPE

Downy mildew is a major disease of grapes throughout the eastern United States. The fungus causes direct yield losses by rotting inflorescences, clusters and shoots. Indirect losses can result from premature defoliation of vines due to foliar infections. This premature defoliation is a serious problem because it predisposes the vine to winter injury. It may take a vineyard several years to fully recover after severe winter injury.

SYMPTOMS

On leaves, young infections are very small, greenish-yellow, translucent spots that are difficult to see. With time the lesions enlarge, appearing on the upper leaf surface as irregular pale-yellow to greenish-yellow spots up to 1/4 inch or more in diameter. On the underside of the leaf, the fungus mycelium (the "downy mildew") can be seen within the border of the lesion as a delicate, dense, white to grayish, cotton-like growth. Infected tissue gradually becomes dark brown, irregular, and brittle. Severely infected leaves eventually turn brown, wither, curl, and drop.

The disease attacks older leaves in late summer and autumn, producing a mosaic of small, angular, yellow to red-brown spots on the upper surface. Lesions commonly form along veins, and the fungus sporulates in these areas on the lower leaf surface during periods of wet weather and high humidity.

On fruit, most infection occurs during 2 distinct periods in the growing season. The first is when berries arᶜ about the size of small peas. When infected at this stage, young berries turn light brown and soft, shatter easily, and under humid conditions are often covered with the downy-like growth of the fungus. Generally, little infection occurs during hot summer months. As nights become cooler in late summer or early fall, the second infection period may develop. Berries infected at this time generally do not turn soft or become covered with the downy growth. Instead, they turn dull green, then dark brown to brownish-purple.

They may wrinkle and shatter easily and, in severe cases, the entire fruit cluster may rot. These infected fruit will never mature normally. On shoots and tendrils, early symptoms appear as water-soaked, shiny depressions on which the dense downy mildew growth appears. Young shoots usually are stunted and become thickened and distorted. Severely infected shoots and tendrils usually die.

Causal Organism

Downy mildew is caused by the fungus Plasmopara viticola. The fungus overwinters in infected leaves on the ground and possibly in diseased shoots. The overwintering spore (oospore) germinates in the spring and produces a different type of spore (sporangium). These sporangia are spread by wind and splashing rain. When plant parts are covered with a film of moisture, the sporangia release small swimming spores, called zoospores. Zoospores, which also are spread by splashing rain, germinate by producing a germ tube that enters the leaf through stomates (tiny pores) on the lower leaf surface.

The optimum temperature for disease development is 64 to 76 degrees F (18 to 25 degrees C). The disease can tolerate a minimum temperature of 54 to 58 degrees F (12 C to 13 degrees C), and a maximum temperature of about 86 degrees F(30 degrees C). Once inside the plant, the fungus grows and spreads through tissues. Infections are usually visible as lesions in about 7-12 days.

At night during periods of high humidity and temperatures above 55 degrees F (13 degrees C), the fungus grows out through the stomates of infected tissue and produces microscopic, branched, tree-like structures (sporangiophores) on the lower leaf surface. More spores (sporangia) are produced on the tips of these tree-like structures. The small sporangiophores and sporangia make up the cottony, downy mildew growth. Sporangia cause secondary infections and are spread by rain.

Control

Any practice that speeds the drying time of leaves and fruit will reduce the potential for infection. Select a planting site where vines are exposed to all-day sun, with good air circulation and soil drainage. Space vines properly in the row, and, if possible, orient the rows to maximize air movement down the row. Sanitation is important. Remove dead leaves and berries from vines and the ground after leaf drop. It may be beneficial to cultivate the vineyard before bud break to cover old berries and other debris with soil.

Cultivation also prevents overwintering spores from reaching developing vines in the spring. To improve air circulation, control weeds and tall grasses in the vineyard and surrounding areas. When pruning, select only strong, healthy, well-coloured canes of the previous year's growth. Practices such as shoot positioning and leaf removal that help to open the canopy for improved air circulation and spray coverage are also very important.

Grape varieties vary greatly in their susceptibility to downy mildew. In general, vinifera (Vitis vinifera) varieties are much more susceptible than American types, and the French hybrids are somewhat intermediate in susceptibility. Cabernet Franc, Cabernet Souvignon, Catawba, Chancellor, Chardonnay, Delaware, Fredonia, Gewurytraminer, Ives, Merlot, Niagra, Pinot Blanc, Pinot Noir, Riesling, Rougeon and Sauvignon Blanc are reported to be highly susecptible to downy mildew.

A good fungicide spray programme is extremely important. Downy mildew can be effectively controlled by properly timed and effective fungicides.

RED STELE ROOT ROT OF STRAWBERRY

Many commercial strawberry cultivars are susceptible to the red stele fungus. This root rot disease has become a serious problem facing strawberry production in the northern two-thirds of the United States. The disease is most destructive in heavy clay soils that are saturated with water during cool weather when the fungus is most active.

The red stele fungus can survive in soil for up to 13 years or longer once it becomes established in the field or garden. Normally, the disease is prevalent only in the lower or poorly drained areas of the planting; however, it may become fairly well distributed over the entire patch, especially during a cool, wet spring.

SYMPTOMS

When plants start wilting and dying in the lower portions of the strawberry planting, the cause is very likely to be red stele. Infected plants are stunted, lose their shiny-green luster, and produce few runners. Younger leaves often have a metallic, blueish-green cast.

Older leaves turn prematurely yellow or red. With the first hot, dry weather of early summer, diseased plants wilt rapidly and die. Diseased plants have very few new roots, when compared with the roots of healthy plants that have thick and bushy roots with many secondary feeding roots. Infected strawberry roots usually appear gray, while the new roots of a healthy plant are yellowish-white.

The most reliable symptom of red stele is found within the roots and may be observed by gently digging up a few plants that are just beginning to wilt, taking care to preserve the root system. Plants with red stele usually have few fine lateral roots so the main fleshy roots have a "rat-tail" appearance. During intermediate stages of disease development, these fleshy roots will be white near the crown of the plant but will show a dark rot progressing upward from the tips.

When the white outer portion of the root just above this rotten zone is peeled off or sliced through, the root core (or stele) will appear to be dark red. It may be necessary to examine several rotting roots before finding a red stele, but this symptom is very distinctive and is diagnostic for the disease. Reddened steles are relatively difficult to find after harvest because most infected roots have died and begun to decay by then.

Causal Organism and Disease Cycle

Red stele is caused by the soilborne fungus Phytophthora fragariae. This fungus is not a natural inhabitant of most agricultural soils but probably is introduced on nursery stock or by the movement of infested soil and runoff water from fields in which the disease occurred previously. P. fragariae is very persistent and can survive in a field for many years once it has become established, even if no strawberries are grown during that time.

The organism that causes red stele of strawberry is not known to cause disease on any other crop, with the possible exception of loganberry. P. fragariae persists in the soil as thick-walled resting spores (oospores). When the soil is moist or wet, some of the oospores germinate and form structures called sporangia, which are filled with the infectious spores of the fungus (zoospores).

These microscopic zoospores are released into the soil when it becomes completely saturated with water (flooded or puddled) and use tail-like structures to swim short distances through water-filled soil pores to the tips of strawberry roots, to which they are chemically attracted. Zoospores also may swim to the soil surface, where surface runoff water may carry them relatively long distances.

Zoospore activity may occur at soil temperatures ranging from about 38 to 77 degrees F (4 to 25 degrees C), but is most significant from 44 to 59 degrees F (7 to 15 degrees C). Thus infection is most likely in the spring and fall. Once zoospores have infected the root tip, the fungus begins to grow up into other parts of the root, causing the characteristic dark rot and red stele symptoms.

New sporangia are formed along the outside of infected root issue and release additional zoospores whenever the soil is saturated, thereby continuing to spread the disease. The fungus produces oospores within infected roots as they begin to rot and die, and these oospores are released into the soil when the roots decay, thus completing the disease cycle.

Control

Since significant production and movement of infective zoospores occurs only during periods when the soil is completely saturated, the key to control is drainage. Strawberries should not be planted in low-lying or heavy soils where water accumulates or is slow to drain.

On marginal soils, planting strawberries on beds raised at least 10 inches high will bring much of the root system above the zone of greatest pathogen activity and the severity of red stele root rot should be significantly reduced. Strawberry varieties highly resistant to red stele should be seriously considered for planting in a marginally drained site or a field in which red stele has been suspected of occurring in the past.

Only resistant varieties should be planted in a field where red stele is known to have caused losses within the last 5 to 10 years. The following junebearing varieties are reported to be resistant to Red Stele: Allstar; Delite; Earliglow; Guardian; Lester; Midway; Redchief; Scott; Sparkle; Sunrise and Surecrop. The everbearing varieties are also reported to be resistant. All "resistant" varieties, however, are resistant only to certain common races of the red stele fungus and can become diseased if exposed to other races of the pathogen.

It is important to minimize the chance of introducing the red stele fungus into a field where it does not already exist.

Buy nursery stock only from a reputable supplier, and take care not to transfer soil on farm implements from an infested field into a clean one. New fungicides active against red stele also help in controlling this disease but are most effective when used in combination with good soil water management practices.

STRAWBERRY LEAF SPOT DISEASES

BLIGHT ON STRAWBERRY LEAVE

Fungal diseases of the leaf may occur as soon as the first

leaves unfold in early spring and continue until dormancy in the late fall. Generally, these diseases do not cause significant economic damage. The primary damage from leaf diseases is a loss of vigour through reduced leaf area. If outbreaks of these leaf diseases become significant, the plants will become weakened resulting in increased susceptibility to root diseases and winter injury.

The three major leaf diseases that are caused by fungi have a similar disease cycle and are controlled in a similar manner. Leaf spot, leaf scorch, and leaf blight are the most common leaf diseases and they all overwinter in infected dead or living leaves. They all produce spores that spread the disease by causing new infections during moist, warm conditions.

LEAF SPOT

Leaf spot is caused by the fungus, Mycosphaerella fragariae. Symptoms of leaf spot first appear as circular, deep purple spots on the upper leaf surface. These spots enlarge and the centres turn grayish to white on older leaves and light brown on young leaves. A definite reddish purple to rusty brown border surrounds the spots. On fruit, superficial black spots may form under moist weather conditions.

The spots form on ripe berries around groups of seeds. The spots are about 1/4 inch in diameter, and there are usually only one or two spots per fruit. However, some fruits may be more severely infected. The fungus overwinters as spores in lesions on leaves.

The fungus produces more spores in spots on the upper and lower leaf surface that spread the disease during early summer. These spores are spread by splashing rain. Middle-aged leaves are most susceptible. Lesions also develop on stems, petioles and runners.

LEAF SCORCH

Leaf scorch is caused by the fungus Diplocarpon earliana. Symptoms of leaf scorch consist of numerous small, irregular, purplish spots or "blotches" that develop on the upper surface of leaves. The centres of the blotches become brownish.

Blotches may coalesce until they nearly cover the leaflet, which then appears purplish to reddish to brown. The fungus overwinters on infected leaves. The fungus produces spore forming structures in the spring on both surfaces of dead leaves. These structures produce spores abundantly in midsummer. In the presence of free water these spores can germinate and infect the plant within 24 hrs. Older and middle-aged leaves are infected more easily than young ones.

LEAF BLIGHT

Leaf blight is caused by the fungus, Phomopsis obscurans. Symptoms of leaf blight infections begin as one to several circular reddish-purple spots on a leaflet. Spots enlarge to V-shaped lesions with a light brown inner zone and dark brown outer zone. Lesions follow major veins progressing inward. The whole leaflet may turn brown.

In severe cases, stolons, fruit trusses and petioles may become infected which may girdle and kill the stem. The fungus overwinters as mycelium or fruiting structures on the old leaves that remain attached to the plant. Spores are spread by rain splash early in the spring. Leaf blight is most destructive to older leaves in the late summer. Petioles, calyxes and fruit may also be infected earlier in the season.

Control

Leaf spot and leaf scorch are controlled most effectively by the use of resistant varieties. The following junebearing varieties are reported to be resistant to both leaf spot and leaf scorch: Allstar; Canoga; Cardinal; Delite; Earliglow; Honeoye; Jewell; Lester; Midway and Redchief. The ever bearing varieties, Tribute and Tristar, are reported to be tolerant to leaf spot and leaf scorch. There are no varieties with reported resistance to leaf blight. These cultural practices should help reduce infection:

- Remove the older and infected leaves from runner plants before setting.
- Take care in spacing runner plants in matted-row culture.

- Plant in light, well-drained soil in a location exposed to all-day sun and good air circulation.
- Control weeds in the planting. Weeds reduce air circulation and increase drying time for leaves.
- Removing infected leaves after harvest (during renovation) is helpful in reducing inoculum and controlling all the leaf diseases.

If leaf diseases are a problem in the planting, fungicides will aid in control.

BOTRYTIS FRUIT ROT "GRAY MOLD"

Many fungi are capable of rotting mature or near-mature fruits of strawberry, raspberry, and blackberry. Under favourable environmental conditions for disease development, serious losses can occur. One of the most serious and common fruit rot diseases is gray mold. The gray mold fungus can affect petals, flower stalks (pedicels), fruit caps, and fruit. In wet, warm seasons, probably no other disease causes a greater loss of flowers and fruit. The disease is most severe during years with prolonged rainy and cloudy periods during bloom or during harvest.

SYMPTOMS

Young blossoms are usually very susceptible to infection. One or several blossoms in a cluster may show blasting (browning and drying) that may extend down the pedicel. Fruit infections usually appear as soft, light brown, rapidly enlarging areas on the fruit. If infected fruits remain on the plant, the berry usually dries up, "mummifies," and becomes covered with a gray, dusty powder, which gives the disease its name "gray mold."

Fruit infection is most severe in well-protected areas of the plant, where the humidity is high and air movement is poor. On strawberry, berries resting on soil or touching another decayed berry or a dead leaf in dense foliage are most commonly affected.

The disease may develop on young green fruits, but fruits become more susceptible as they mature. Usually, the disease is not detected until fruits are mature at harvest time. After picking, mature fruits are extremely susceptible to gray mold, especially if bruised. During picking, the handling of infected fruit will spread the fungus to healthy ones. Under favourable conditions for disease development, healthy berries may become a rotted mass within 48 hours after picking.

Causal Organism

Gray mold is caused by the fungus Botrytis cinerea. The fungus is capable of infecting a great number of different plants. The disease cycle is very similar for both strawberries and brambles.

The fungus overwinters as minute, black, fungus bodies (sclerotia) or as mycelium in plant debris, such as dead strawberry or raspberry leaves. Recent research has shown that nearly all of the overwintering inoculum in strawberry plantings comes from mycelium in dead strawberry leaves within the row or planting.

In early spring, the mycelium becomes active and produces large numbers of microscopic spores (conidia) on the surface of old plant (leaf) debris in the row. Spores are spread by wind throughout the planting where they are deposited on blossoms and fruits. They germinate when a film of moisture is present and infection can occur within a few hours. Temperatures between 70 and 80 degrees F (20 to 27 degrees C) and free moisture on the foliage from rain, dew, fog, or irrigation water are ideal conditions for disease development.

The disease can develop at lower temperatures if foliage remains wet for long periods. Strawberries and raspberries are susceptible to Botrytis during bloom and again as fruits ripen. Recent research indicates that most fruit infection actually occurs during bloom; however, symptoms usually do not develop until close to harvest. During bloom, the fungus colonizes healthy or senescing flower parts, often turning the blossoms brown.

These blossom infections establish the fungus within the receptacle of the young fruit as a "latent" or "quiescent" infection. The fungus generally remains latent in developing (green) fruit until the fruit starts to mature, at which time the fungus becomes active and symptoms (rot) appear.

Thus, the most critical period for applying fungicides to control gray mold is during bloom. This is an important point to remember when considering fungicide applications for controlling this disease.

Control

Select a planting site with good soil drainage and air circulation. Plants should be exposed to direct sunlight. Plant rows with the direction of the prevailing wind to promote faster drying of foliage and fruit. A good layer of straw mulch (or other material) between the rows or around the plants aids greatly in controlling fruit rots. The mulch acts as a barrier that reduces fruit contact with the soil. Proper spacing of plants and timing of fertilizer applications are also important.

Excessive applications of nitrogen fertilizer, especially in the spring before harvest, can produce excessive amounts of dense foliage. Shading of berries by thick foliage prevents rapid drying of the fruit during wet periods and creates ideal conditions for disease development. Good weed control is very important. Weeds prevent air movement in the plant canopy. This slows drying time of flowers and fruits and increases the chances for infection. Pick fruit frequently and early in the day as soon as plants are dry.

Cull out all diseased berries but do not leave them in the field. Handle berries with care to avoid bruising. Refrigerate fruit promptly at 32 to 50 degrees F to check gray mold. Fungicides are an important disease management tool in commercial plantings, but are generally not effective unless they are timed properly and used in conjunction with the above mentioned cultural practices. Homeowners are encouraged to emphasize the use of cultural practices in order to avoid the use of fungicides.

POWDERY MILDEW OF GRAPE

Powdery mildew is an important disease of grapes

throughout. The disease generally is considered less economically important than black rot or downy mildew. However, uncontrolled, the disease can be devastating on susceptible varieties under the proper environmental conditions.

Unlike black rot and downy mildew, the powdery mildew fungus does not require free water on the plant tissue surface to infect. Powdery mildew can result in reduced vine growth, yield, fruit quality, and winter hardiness. Varieties of Vitis vinifera and its hybrids generally are much more susceptible than American varieties.

SYMPTOMS

The powdery mildew fungus can infect all green tissues of the vine. Small, white or grayish-white patches of fungal growth appear on the upper or lower leaf surface. These patches usually enlarge until the entire upper leaf surface has a powdery, white to gray coating.

The patches may remain limited throughout most of the season. Severely affected leaves may curl upward during hot, dry weather. Expanding leaves that are infected may become distorted and stunted. On young shoots, infections are more likely to be limited, and they appear as dark-brown to black patches that remain as dark patches on the surface of dormant canes. If blossom clusters are affected, the flowers may wither and drop without setting fruit. Infections on cluster stems often go unnoticed, but can be very damaging. Infected cluster stems may wither and dry up, resulting in berry drop (shelling). Affected berries may have spots on the surface similar to those on the leaves, or the entire berry may be covered with the white, powdery growth.

Infected berries often are misshapen or have rusty spots on the surface. Severely affected fruit often split open. When berries of purple or red cultivars are infected as they begin to ripen, they fail to colour properly and have a blotchy appearance at harvest. Berries are susceptible to infection until their sugar content (0Brix) reaches about 8 per cent. Late in the season, many black specks may develop on the surface of infected areas. These are the sexual fruiting bodies (cleistothecia) of the fungus.

Causal Organism and Disease Cycle

Powdery mildew is caused by the fungus Uncinula necator. It was previously thought that the powdery mildew fungus overwintered inside dormant buds of the grapevine. Recent research has shown that almost all overwintering inoculum in comes from cleistothecia, which are fungal fruiting bodies that overwinter primarily in bark crevices on the grapevine. In the spring, airborne spores (ascospores) released from the cleistothecia are the primary inoculum for powdery mildew infections.

Ascospore discharge from cleistothecia is initiated if 0.1 inch of rain occurs with an average temperature of 50 degrees F. Most mature ascospores are discharged within 4-8 hours. Ascospores are carried by wind. They germinate on any green surface on the developing vine, and enter the plant resulting in primary infections. The fungus produces another type of spore (conidia) over the infected area after 6-8 days. The conidia and fungus mycelia on which they are formed give the powdery or dusty appearance to infected plant parts. The conidia serve as "secondary inoculum" for powdery mildew infection throughout the remainder of the growing season.

It is important to note that a primary infection caused by one ascospore can result in the production of hundreds of thousands of conidia, each of which is capable of causing secondary infections that spread the disease.

Temperatures of 68-77 degrees F are optimal for infection and disease development, although infection can occur from 59-90 degrees F.

Temperatures above 95 degrees F inhibit germination of conidia and above 104 degrees F they are killed. High relative humidity is conducive to production of conidia. Atmospheric moisture in the 40-100 per cent relative humidity range is sufficient for germination of conidia and infection.

Free moisture, especially rainfall, is detrimental to survival of conidia. This is in direct contrast to most other grape diseases such as black rot and downy mildew that require free water on the plant surface before the fungus spores can germinate and infect. Low, diffuse light seems to favour powdery mildew development. Under optimal conditions, the time from infection to production of conidia is about 7 days.

It is important to remember that powdery mildew can be a serious problem in drier growing seasons when it is too dry for other diseases such as black rot or downy mildew to develop. Cleistothecia are formed on the surface of infected plant parts in late fall. Many of them are washed into bark crevices on the vine trunk where they overwinter to initiate primary infections during the next growing season.

Control

Select an open planting site with direct sunlight. Plant rows in the direction of the prevailing wind in order to promote good air circulation and faster drying of foliage and fruit. Prune and train vines properly in such as way as to reduce shading and increase air circulation. Varieties differ greatly in their susceptibility to powdery mildew.

Cabernet Franc, Cabernet Sauvigon, Chancellor, Chardonnay, Chelois, Gewurztraminer, Merlot, Pinot blanc, Pinot noir, Riesling, Rosette, Rougeon, Sauvignon blanc, Seyval, Vidal 256 and Vignoles are all highly susceptible. On susceptible varieties, control is based on the use of properly timed applications of effective fungicides. Early season (prebloom through bloom) control of primary infections caused by ascospores must be emphasized.

BACTERIAL SPOT OF STONE FRUITS

Bacterial spot affects peaches, nectarines, apricots, plums, prunes, and cherries. The disease is widespread throughout all fruit growing states east of the Rocky Mountains. Bacterial spot can affect leaves, twigs, and fruit. Severe infection results in reduced fruit quality and yield. Fruit infection is most serious on late-matùring varieties.

If proper environmental conditions occur, up to 50 per cent or more of the fruit of susceptible varieties may have to be discarded. Defoliation by bacterial spot may weaken the tree, predisposing it to winter injury and attack by other pathogens. The disease is usually more severe where soils are light and low in fertility. Vigorous trees are usually less susceptible to the disease than devitalized, neglected trees.

SYMPTOMS

Leaves

Small (l/25 to l/5 inch) spots form in the leaves. Spots are irregular to angular and have a deep purple to rusty-brown or black colour. In time, the centres dry and tear away leaving ragged "shot-holes." When several spots merge, the leaf may appear scorched, blighted, or ragged. Badly infected leaves may turn yellow and drop early. Early defoliation is most common on trees deficient in nitrogen or where the disease is further complicated by pesticide injury. Nitrogen deficiency can cause leaf symptoms that are very similar to those of bacterial spot. Care must be taken to avoid an incorrect diagnosis.

In both cases, the leaves of affected trees turn yellow and drop prematurely. With bacterial spot, symptoms are most likely to appear at the tip of the leaf but with nitrogen deficiency, symptoms are usually most evident along the midrib. Leaf tissues surrounding shot-holes caused by a nitrogen deficiency are more likely to have a reddish colour than when bacterial spot is involved.

Fruit

Small, round olive-brown to black spots form on the fruit. They are usually sunken and frequently surrounded by a water-soaked margin. On peaches, spots usually form on the side exposed to the sun. Spots may slowly enlarge and merge to cover large irregular areas on the fruit. On some varieties the spots may exude a yellowish gum after rainy periods. Skin cracking and pitting may occur near the spots during fruit enlargement. Fruit infected at an early stage of development are usually the most malformed.

On plums, symptoms are different than on peaches. Large, sunken, black spots form on some varieties; on others, small pit-like lesions are common.

Twigs

On peaches, two distinct types of cankers damage twigs. "Spring cankers" develop on young twigs produced the previous summer. Spring cankers first appear as water-soaked, slightly darkened blisters about the time the first leaves appear. If these cankers encircle the twig, it will die. As the season progresses, the tissues over the blister-like lesions rupture and bacteria are released.

These bacteria can be spread by windblown or splashing rain and can result in new infections. In time, spring cankers heal and become inactive. "Summer cankers" develop on green twigs of the current seasons growth.

They usually occur later in the summer after leaf spots are evident. At first, they are water-soaked, dark purplish spots. In time, they enlarge, turn brown to purple-black, become slightly sunken, and are round to elliptical with water-soaked margins. On certain plum and apricot varieties, twig cankers may continue to develop in two- and three-year old twigs. If the cankers are deep-seated, they can deform or kill the twigs.

Causal Organism

Bacterial spot is caused by the bacterium Xanthomonas pruni. The bacteria overwinter in twigs that are infected late in the season about the time leaves are shed. The following spring, when environmental conditions are favourable, bacteria ooze out onto the surface of these twigs. The bacteria are then spread by windblown or splashing rain and can result in new infections throughout the growing season.

The bacteria come in contact with healthy leaves, fruit, and current-year twigs and enter the tissues through stomata or lenticels when surface moisture is present. Once inside healthy tissues, the bacteria multiply and disease develops. Warm temperatures (70-85 degrees F, 21-29 degrees C) with light rains, heavy dews, and windy weather are most conducive for disease development and spread. The disease makes little progress when weather is hot and dry.

Control

- When planning an orchard, avoid low-lying or shaded sites with poor air circulation and soil drainage. Any practice that promotes faster drying of fruit and foliage will help reduce the risk of infection. Destroy nearby wild or neglected stone fruits (Prunus species). Buy and plant only vigorous, disease-free fruit trees from a reputable nursery.
- Prune trees annually to allow for better air circulation and to maintain tree vigour. If possible, prune during dry weather in the latter half of the dormant season.
- Select peach varieties with resistance to bacteria spot. The following varieties are somewhat resistant: Belle of Georgia, Biscoe, Candor, Comanche, Garnet Beauty, Harbrite, Harken, Late Sunhaven, Loring, Madison, Norman, Pekin, Raritan Rose, Redhaven, Redskin and Sunhaven. These varieties are very susceptible: Babygold S, Blake, Elberta, Halehaven, Jersey Queen, Jerseyland, July Elberta, J.H. Hale, Kalhaven, Rio-Oso-Gem, Suncling, Suncrest, and Sunhigh.
- Fertilize where needed to maintain vigorous but not excessive shoot growth.
- Spray applications. At present, no spray programme is completely effective for controlling bacterial spot. In the home orchard, spraying for bacterial spot is not considered practical.

SCAB OF PEACH, NECTARINE, PLUM, AND APRICOT

Scab occurs throughout the Midwest, wherever peaches, nectarines, plums, and apricots are grown. The disease affects fruit, leaves, and young green twigs. Scab is most common in home orchards where fungicide spray programmes are not practiced. The general use of fungicides by commercial growers has greatly reduced losses from scab. The principle loss from scab is the unsightly spotting of the fruit skin, but premature defoliation and a dieback of infected twigs can also occur. Losses are generally greater on peaches than on nectarines, plums, and apricots. Wet conditions during spring and early summer after petal fall are necessary for a severe attack by the scab fungus. The disease is usually more serious in low-lying, moist, shady areas where air movement is poor.

SympToms

Symptoms first appear on fruit when they are half formed to nearly fullgrown, about six to seven weeks after petal fall. Small, round, olive-green spots about 1/16-1/8 inch in diameter develop on the fruit. Spots are most common near the stem end on the side of the fruit exposed to the sun. Spots are superficial and slowly enlarge.

They may merge to form large, irregular blotches that turn velvety, dark olive-green or black. Severely infected fruit may be stunted, become misshapen, or crack open. When fruits crack open, they are often invaded by other fungi that rot the fruit. Scabby fruit may also drop prematurely and do not ship or store well. Leaves may also be infected by scab. Small (1/4 inch), round, and yellowish-green to yellowish-brown spots develop on the underside of the leaf. Diseased leaf tissue may dry up and drop out, leaving "shot-holes." If the season is wet, scab-infected leaves usually drop early.

Causal Organism

Scab is caused by the fungus Cladosporium carpophilum. The scab fungus overwinters in twigs infected the previous year. During spring and summer, large numbers of microscopic spores (conidia) are formed on the twig lesions. At 100 per cent relative humidity, 20 to 30 hours are required for abundant sporulation in twig lesions. The spores remain firmly attached to the twig until they are moistened.

When moistened, they are spread by splashing rain or windblown mist to developing fruit, twigs, and leaves. Spores are produced in the greatest numbers about two to six weeks after the shucksplit stage of development during warm, wet weather. Spore germination and fungus growth is most rapid at 65-75 degrees F. If weather conditions are favourable, infection begins to occur at about shuckfall. The fruit remain susceptible until harvested. Forty to 70 days elapse from the time the spore lands on the fruit until the disease is visible. Thus, the disease is usually not observed until the fruit are well grown. Spores from the fruit reinfect the twigs and leaves, completing the disease cycle.

Control

- When planting an orchard, avoid low-lying or shaded sites with poor air circulation and soil drainage.
- Any practice that promotes faster drying of fruit and foliage will help reduce the risk of infection.
- Destroy nearby wild or neglected peaches, nectarines, plums, and apricots.
- Remove and destroy scab-blighted twigs before growth starts in early spring.
- Backyard growers should remember that infections are generally superficial. If infection is not severe, fruit quality is little affected. Peeling infected fruits make them quite acceptable for processing (canning or freezing) or eating fresh.

CHERRY LEAF SPOT

Cherry leaf spot is one of the most serious diseases of both sweet and sour cherries in the Midwest. The disease mainly affects the leaves, but lesions may also appear on fruit, petioles, and fruit stems (pedicels). Diseased leaves drop prematurely, and severely affected trees may be defoliated by mid-summer. Early and repeated defoliation can result in:

- Dwarfed and unevenly ripened fruit with poor taste;
- Devitalized trees that are more susceptible to winter injury;
- Death of fruit spurs;
- A reduction of fruit set and size;
- Small and weak fruit buds;

- Reduced fruit growth;
- Eventual death of the tree.

SYMPTOMS

During the latter part of May and the first half of June, small circular purple spots appear on the upper surface of the leaf. Spots gradually enlarge to about 1/4 inch in diameter and turn reddish-brown. Lesions may emerge to produce large, irregular spots. Whitish-pink masses of sticky spores (conidia) form within the spots on the undersides of infected leaves during periods of damp weather.

After six to eight weeks, the centres of the spots may dry up and drop out, giving a "shot-hole" appearance. The "shot-hole" effect is more common on sour than on sweet cherries. The most conspicuous symptom, especially on sour cherries, is the golden yellowing of older infected leaves before they drop off. Although this symptom does not occur every season, the spotting of infected leaves is always visible. Spots similar to those on the leaves may also form on leaf petioles and fruit pedicels, causing fruit to ripen unevenly. Spots usually do not form on fruit.

Causal Organism

Cherry leaf spot is caused by the fungus, Blumeriella jaapii (previously called Coccomyces hiemalis). The fungus overwinters in dead leaves on the ground. In early spring (about petal fall), fungal fruiting bodies called apothecia develop in these leaves.

Spores (ascospores) are produced in the apothecia and are forcibly discharged during rainy periods for about six to eight weeks, starting at petal fall.

The optimal temperatures for ascospore discharge are 61 degrees F (16 degrees C) and higher. Very few ascospores are discharged at temperatures below 46 degrees F (8 degrees C). These ascospores are spread by wind or splashing rain drops to the green, healthy leaves and serve as primary inoculum for disease. The ascospores stick to the leaf surface, germinate in a film of water, and within several hours at the proper temperature, penetrate the leaf through stomata (natural openings) on the underside of the leaf. The small purple spots soon appear on the upper surface.

Incubation time, from fungus penetration to the appearance of the spots, varies with temperature. Under damp conditions and with temperatures between 60 and 68 degrees F (15-20 degrees C), the period may be as short as five days. When rain and dews are absent and at lower temperatures, as long as 15 days may be required before symptoms appear. Once lesions have developed, masses of secondary or summer spores (conidia) are formed from the slightly concave eruptions (acervuli) on the underside of the leaf. This mass of conidia provides the white appearance to the underleaf lesions.

Conidia are spread to other leaves by splashing raindrops and are capable of causing new infections (each producing thousands of additional conidia) under the temperature and wetness conditions listed in Table. Serious leaf spot damage usually occurs in years with numerous rainy periods throughout late spring and summer, when repeated secondary infection cycles allow the disease to snowball into an epidemic.

Control

- Collect and destroy the fallen cherry leaves in late autumn. The fungus overwinters in these leaves. This practice should be quite beneficial for backyard growers with one or a few trees, but is generally not considered practical for large commercial plantings.
- Select a planting sight that is always exposed to direct sunlight and has good air circulation and soil drainage. Proper pruning to open the canopy will increase sunlight penetration and air circulation. Any practice that promotes faster drying of leaves will reduce the risk of infection.
- Especially in commercial plantings, the disease is controlled primarily with fungicide sprays.

SOOTY BLOTCH AND FLY SPECK OF APPLE

Sooty blotch and fly speck are two different diseases.

Both diseases are widespread throughout the Midwest, and commonly occur together on the same fruit.

Sooty blotch and fly speck cause a discolouration or blemish of near-mature fruit. The discolouration is superficial, and while neither disease actually damages the fruit, the presence of disease reduces the grade and market value of the fruit. Although all apple varieties are susceptible to infection by both fungi, symptoms are most severe on yellow or light-coloured varieties such as Golden Delicious or Grimes. Both diseases are most common during years with a cool, wet spring, rains in late summer and low temperatures in early fall.

SYMPTOMS

Brown to dull black, sooty blotches with an indefinite outline form on the fruit surface. Blotches may be 1/4 inch in diameter or larger. The blotches may coalesce to cover practically the entire fruit. The sooty blotch fungus is restricted to the outer surface of the fruit, and in many cases the blotches can be easily rubbed off.

However, if infection occurs early in the season, we may need to rub or bleach the fruit vigorously to remove it. Groups of 6 to 50 or more black and shiny round dots that resemble fly excreta appear on the surface of the fruit. The individual "fly specks" are clearly separated and can be easily distinguished from sooty blotch. Like sooty blotch, fly speck infections are superficial; however, they are usually harder to rub off than sooty blotch.

Causal Organism

The fungus Gloeodes pomigena causes sooty blotch, and Zygophiala jamaicensis causes fly speck. Both fungi overwinter on twigs of various wild woody plants, especially wild blackberry and raspberry canes. Both fungi require free water on the fruit surface to infect.

SOOTY BLOTCH

The "sooty blotch" or "smudge" appearance on affected

fruit results from the presence of hundreds of minute, dark fungal fruiting bodies (pycnidia) that are interconnected by a mass of loose, interwoven dark hyphae (fungal filaments).

In spring, pycnidia on wild plants produce large numbers of spores (conidia) that ooze out and collect in a gelatinous mass. The conidia are then spread by water splash or wind blown mists into orchards from late May or early June until fall. The fungus first affects apple twigs, and then secondary colonies are initiated on the fruit.

Cool, humid weather is essential for disease development. The disease does not develop when temperatures reach 85 degrees F. When May and June are cool and moist and are followed by a hot July and August, sooty blotch symptoms often do not appear for two to three months. The disease is absent or rare when hot, dry weather prevails until close to harvest time.

The disease is most severe when cool, rainy weather in the spring is coupled with late summer rains and low temperatures in early fall.

Under ideal conditions, the incubation period from the time the fungus reaches the fruit to the appearance of symptoms may be as short as five days. In the orchard, the incubation period usually lasts three to four weeks on fruits that are 42 to 45 days old.

FLY SPECK

The individual "fly specks" are sexual fruiting bodies (ascocarps) of the fungus. Starting in late spring, the fungus produces spores on wild hosts. These spores are carried by wind into the orchard. When spores come into contact with the fruit under the proper environmental conditions, they germinate and infect. Symptoms can develop within 15 days under favourable environmental conditions (65 degrees F or 18 degrees C).

Control

- Select an orchard site that always has full sunlight, good air circulation, and good soil (water) drainage.

- Prune trees annually to an open centre for maximum air circulation. Both diseases are most prevalent in the damp, low, shaded areas of the orchard. Any practice that opens up the trees to greater air movement and promotes faster drying greatly aids in control.
- Remove or destroy nearby wild or neglected apple trees.
- Backyard growers should remember that the disease is superficial and rarely affects the quality of the fruit. Removal of the fungus by washing, rubbing, or peeling the fruit results in fruit that is acceptable for cooking or eating fresh.
- Especially in commercial plantings, fungicide sprays are important for controlling these diseases.

WHITE ROT AND BOTRYOSPHAERIA CANKER OF APPLE

White rot of apple fruits is also referred to as "Bot Rot" or Botryosphaeria rot. The fungus that causes fruit rot can also cause a canker on limbs and other above-ground woody portions of the tree. The canker phase of the disease is most severe in trees weakened by drought, winter injury, sunscald, poor pruning, low or unbalanced nutrition, and other plant diseases. The fruit rot phase can be sporadic in appearance, being serious one season and difficult to find in the following season. The Botryosphaeria fungus attacks a wide range of woody plants that are common. Dutchess, Golden Delicious, Grimes Golden, Gallia Beauty, Rome, and Yellow Transparent apple varieties are all very susceptible to fruit rot. Jonathan and Red Delicious are less likely to be affected than other varieties.

SYMPTOMS

Fruit

At first, small, reddish-brown spots appear around the lenticels. The spots enlarge and become slightly depressed.

On yellow-skinned varieties, these spots may be bordered by one or more red "halo" rings. Spots on red-skinned apples often become bleached.

The tissue under the spots is soft and egg shaped, with the long axis parallel to that of the core. Several spots may merge to involve all or much of the fruit. As the rot progresses, the skin colour becomes dark brown and superficially resembles black rot, except that with black rot the decayed tissue is firm, instead of soft and mushy.

Beads of exudate appear on the surface of fruits completely rotted by white rot. Small, black fruiting bodies (pycnidia) that are filled with spores (conidia) develop on the surface of rotted fruits during warm, moist conditions. Mature fruits are most susceptible to the disease. Apple fruits often become infected from mid-summer on without showing external symptoms.

Thus, fruit may be infected without symptoms appearing in the orchard. Disease development is checked when these apples are placed in cold storage. However, when they are removed from cold storage an internal rot develops and the fruit may deteriorate very rapidly at room temperature.

Twigs and Limbs

Small, circular spots or "blisters" appear on the twigs in June and July. The spots enlarge, become somewhat sunken, and fill with a watery fluid. The fungus may grow rapidly through the tissues to form slightly sunken, dark-coloured cankers that may extend to the cambium on very susceptible apple varieties.

Under favourable conditions, several cankers may fuse to girdle and kill large limbs. On older cankers, the outer bark becomes tan to orange and papery, and the margins of the canker crack and fissure.

The outer bark sloughs off and the underlying tissue appears slimy. In the fall, twig and limb cankers stop growing and may split along the edges. Rings of small, black, spore-producing bodies (pycnidia and perithecia) are formed on the surface of the cankers or under the papery outer bark. The following spring, a canker may resume growth or be corked off and become inactive.

CASUAL ORGANISM AND DISEASE CYCLE

Botryosphaeria canker and fruit rot (white rot) is caused by the fungus Botryosphaeria dothidea. The fungus overwinters as black pycnidia and perithecia in a wart-like stroma on living and dead cankered limbs and in rotted fruits. The fungus is also commonly found on fire-blighted twigs or cankers. Wounds or breaks in the epidermis are necessary for the fungus to penetrate. Spores (ascospores) are forcibly discharged from perithecia during spring rains.

Another type of spore (conidia) is produced within pycnidia and ooze out in tremendous numbers. They are then washed and rain-splashed to other parts of the tree throughout the summer.

Apple fruits may become infected fairly early in the season, but rotting does not develop much until the fruit is almost mature. At temperatures above 75 degrees F (24 degrees C), mature fruit may rot completely within a few days after infection. The development of Botryosphaeria canker and fruit rot is favoured by any condition that reduces tree vigour.

Control

Control of white rot is best achieved through an integrated programme of cultural practices and chemical control measures.

- Sanitation is critical for effective control. Piles of prunings are an important source of inoculum and should be removed from the perimeter of the orchard or burned. Prunings can be left on the orchard floor if they are chopped with a flail mower, which removes much of the bark and allows them to decompose faster. Removal of mummified apples and pruning out dead wood in the tree are important for reducing the inoculum within the tree. Pruning out current-season shoots infected with fire blight is also important, because they can be colonized and serve as an inoculum source during the same growing season.

- Any practice that helps to maintain trees in a healthy, vigorous condition is critical for controlling the canker phase of the disease. Cankers generally develop only on stressed or weakened trees. Prune trees annually and maintain a balanced fertility programme based on soil and foliar nutrient analysis. Cankers generally develop rapidly on winter-injured trees.
- The use of fungicides combined with good sanitation is beneficial for controlling the fruit rot phase of the disease. Fungicides are not effective for controlling the canker phase of the disease on weakened trees.

RUST OF APPLE

Three rust diseases occur on apple and crabapple. All are caused by different species of the fungus Gymnosporangium and have various junipers and red cedars (juniperus species) as an alternate host. Apples are generally most susceptible to infection by the rust fungi during the period from early bloom until about 30 days after bloom.

SYMPTOMS

Cedar-Apple Rust on Apple

On leaves, pale yellow spots appear on the upper surface during May or June. Spots are up to 1/4 inch in diameter, turn orange with time, and often have a reddish border. Small black fungal bodies (pycnia) form within the spots and may exude an orange fluid. In time, yellow spots develop on the underside of the leaf. These spots thicken, and during late spring and early summer a number of small, orange-yellow tubular projections (aecia) appear. These develop into open, cylindrical tubes that split toward the base into narrow strips and curl backward.

Infected leaves may turn yellow and drop. Defoliation of rusted leaves is most common in dry summers. On fruit, similar yellow-orange spots appear, usually at or near the calyx end. These spots usually occur on immature fruit and are much larger than the spots on leaves (up to 3/4 inch in diameter). The light green colour of the young fruit becomes a darker green around the infected area. The tube-like aecia may form on the slightly raised fruit lesions. Infected fruits are often stunted and misshapen, and may drop early.

Cedar- Quince Rust on Apple

Cedar-quince rust only affects the fruit of apples. Infected fruit become puckered at the blossom end and later develop a sunken, dark green area. The flesh under the sunken, dark green area becomes brown and spongy. The formation of pycnia and aecia on infected fruit is rare. Apples are susceptible to cedar-quince rust during the period from early bloom through third cover.

Cedar-Hawthorn Rust on Apple Leaf spots similar to those caused by cedar-apple rust develop on apple and crabapple. Larger, gray to brown spots form on leaves of hawthorn. Few aecia form on apple and crabapple. Fruit infection on apple is rare. Defoliation and deformation of fruits and twigs may occur on hawthorns.

Causal Organisms

Cedar-apple rust is caused by the fungus Gymnosporangium juniperi-virginianae, cedar-quince rust by Gymnosporangium clavipes, and cedar-hawthorn rust by Gymnosporangium globosum. The disease cycle for all three rusts is essentially the same. The disease cycle for cedar-apple rust will be presented. The fungus overwinters as mycelium in galls in juniper.

Large yellow to orange gelatinous sporehorns are formed on the galls in the spring and spores (teliospores) are produced. Each teliospore germinates and produces four to eight sporidia or basidiosproes.

As sporehorns begin to dry, the sporidia are forcibly discharged into the air and carried by wind to nearby apple leaves, fruits, and twigs. About 30 days after apples have bloomed, the sporehorns have discharged all their spores and most apple leaves are no longer susceptible. Within five or six hours after landing on the leaf, sporidia become attached to the surface, germinate, and penetrate the cuticle and upper leaf surface. After 10 to 14 days, the yellow spots develop on the upper leaf surface. The orange to black pycnia develop in the spots and several weeks later, the aecia form on the underleaf surface.

The aecia produce another type of spore (aeciospores) that are carried by wind to junipers. When the aeciospores contact a juniper twig, they become firmly attached and germinate in warm, moist weather of late summer or early fall and penetrate the twig. A young, pea-size, greenish-brown gall develops. The gall enlarges the following year, but does not produce sporehorns with teliospores until the second spring. The complete disease cycle requires almost two years.

Control

- Grow resistant or immune apples, crabapples, and junipers. When buying trees, check with the nursery about rust resistance.
- Destroy nearby, worthless or wild junipers infected with rust galls.
- Where rusts are a problem, follow a recommended fungicide spray programme.

BOTRYTIS BUNCH ROT OR GRAY MOLD OF GRAPE

Botrytis bunch rot is caused by the fungus Botrytis cinerea. This fungus is very common in nature, and causes diseases on a variety of unrelated crops. Bunch rot can cause serious losses on highly susceptible grape varieties. Although berries of all grape varieties are susceptible to bunch rot, losses generally are greater on tight-clustered varieties of Vitis vinifera and French Hybrids. Losses result from the rotting of berries in the field or in storage.

SYMPTOMS

Infection of ripe berries is the most common and

destructive phase of this disease. Infected berries first appear soft and watery.

The berries of white cultivars become brown and shriveled, and those of purple cultivars develop a reddish colour. Under high relative humidity and moisture, infected berries usually become covered with a gray growth of fungus mycelium. One or a few berries within the bunch or the entire bunch may be affected.

Generally, healthy berries touching infected berries will become infected. Rotted berries generally shrivel with time and drop to the ground as hard mummies. The fungus also can cause a blossom blight that can result in significant crop loss early in the season. Although uncommon, leaf infections also occur, but appear to be of no economic importance. Leaf infection begins as dull, green spots, commonly surrounded by a vein. The spots rapidly become necrotic lesions.

Casual Organism and Disease Cycle

Botrytis bunch rot is caused by the fungus Botrytis cinerea. The fungus over-winters in grape mummies, dead grape tissues, and other organic debris in and around the vineyard, as well as on a multitude of alternate plant hosts. Because of its wide host range, growers always should assume that the fungus is present in the vineyard. In spring, the fungus germinates from small, dark, hard resting structures known as sclerotia.

The fungus then produces spores (conidia) that spread the disease. These spores are produced throughout the growing season. As blooms die, the spores germinate and colonize dead flower parts. Using the dead tissue as a food base, the fungus invades living tissue. After penetrating the berry, the fungus may remain dormant until the fruit sugar content increases and the acid content decreases to a level that supports fungus growth.

Symptoms then develop readily under warm, moist conditions. Berries that escape bloom-time infection may become infected at or near harvest under favourable environmental conditions.

Any wound on the berry provides an excellent infection site for the fungus even in the absence of favourable environmental conditions. Birds, insects, hail, and powdery mildew are common causes of wounds. Swelling during ripening in tightly packed clusters causes pressure that also can rupture the berries.

Wet and humid conditions around the berries and leaves greatly enhance disease development. The longer wet conditions persist, the greater the probability of infection, even to undamaged berries. Warmer temperatures also favour infection. At 54 to 75 degrees F, infection occurs in 12 to 24 hours, while at 37 degrees F, 60 to 72 hours are required.

Control

- Promote good air circulation and light penetration by proper pruning, controlling weeds and suckers, and positioning or removing shoots for uniform leaf development. Where possible, rows should be planted in the direction of the prevailing wind. Good air circulation and light penetration promote faster drying of plant parts and reduce the risk of disease. Removal of leaves around clusters on mid- or low-wire cordon-trained vines before bunch closing has been shown to reduce losses caused by Botrytis in New York and California vineyards, due to improved air circulation and improved spray penetration and coverage.
- Prevent wounding by controlling insects, birds, and other grape diseases.
- Growth regulators that lengthen the rachis and separate the berries in tight-clustered cultivars can reduce the damage from berries being crushed within the cluster; thus, reducing infection and spread of Botrytis.
- In commercial vineyards, effective fungicides applied at appropriate times during the growing season provide significant control.

VERTICILLIUM WILT OF RASPBERRY

Verticillium wilt is one of the most serious diseases of

raspberry. This disease is caused by a soil-borne fungus and reduces raspberry yields by wilting, stunting, and eventually killing the fruiting cane or the entire plant. The disease is usually more severe in black than in red raspberries. Blackberries are also susceptible to the disease, but seldom suffer severe losses.

Verticillium wilt is usually a cool-weather disease and is most severe in poorly drained soils and following cold, wet springs. The appearance of symptoms on new canes frequently coincides with water stress caused by hot, dry mid-summer weather.

SYMPTOMS

The symptoms usually appear on black raspberries in June or early July, and on red raspberries about a month later. The lower leaves of diseased plants may at first appear to have a dull green cast as compared to the bright green of normal leaves.

Starting at the base of the cane and progressing upward, leaves wilt, turn yellow and drop. Eventually, the cane may be completely defoliated except for a few leaves at the top. Black raspberry canes may exhibit a blue or purple streak from the soil line extending upward to varying heights.

This streak is often not present or difficult to detect on red raspberries. The final effects of the disease are observed on fruiting canes that were infected the year before. In the spring, many of the diseased canes will be dead. Others will be poorly developed and have shriveled buds. The new leaves are usually yellow and stunted. Infected canes may die before fruit matures, resulting in withered, small, and tasteless fruit.

Casual Organism

Verticillium wilt is caused by the fungus, Verticillium albo-atrum. It is a very common soilborne fungus, and has been reported to cause wilt on more than 160 different kinds of plants including strawberries, eggplant, tomatoes, potatoes, stone fruits and peppers.

The Verticillium fungus overwinters in the soil and plant

debris as dormant mycelium or black, speck-sized bodies called microsclerotia. The fungus can survive in the soil for many years. When conditions are favourable, the microsclerotia germinate and produce threadlike fungus filaments (hyphae).

These hyphae can penetrate the root directly; but invasion is aided by breaks or wounds in the roots. Once inside the root, the fungus grows into the water-conducting tissue (xylem). The destruction of water-conducting tissues prevents the movement of water from the roots to the rest of the plant; thus, the plant eventually dies. The fungus produces microslerotia in infected tissues. When these tissues die and are returned to the soil, the disease cycle is completed.

Control

Applications of fungicides are ineffective in control. Soil fumigation has provided excellent control in some locations but is generally very expensive. Reintroduction of the pathogen into fumigated soils, accompanied by a rapid buildup in pathogen populations, is a major concern with using soil fumigation. Rotations (3- to 4-year) with nonsusceptible crops have been recommended for control in Canada but were not effective in California.

Only disease-free nursery stock from fields known to be free of Verticillium should be used to establish new plantings. Satisfactory resistance in commercial raspberry cultivars is not available. It is generally recommended that raspberries not be replanted in an area where the disease has been a problem. If they are replanted in an infested site, soil fumigation should be considered.

BLACK ROOT ROT OF STRAWBERRY

Black root rot is a serious and common problem of strawberries. The term "Black root rot" is the general name for several root disorders that produce similar sym toms. The disorders are not clearly understood and are generally referred to as a root-rot complex. For this reason, it is difficult to discuss black root rot as we do other diseases which usually have a specific cause. Black root rot has been found in every strawberry growing area of the United States, and a considerable incidence of black root rot has been observed in recent years.

Symptoms

Black root rot is most common in fields with a long history of strawberry production. Symptoms begin with some plants in a field showing reduced vigour, often in low or wet spots or in portions of the field where the soil has become compacted. This decline in vigour usually begins during the first fruiting year. The symptoms are most apparent the last couple of weeks before harvest. Although severely affected plants may die before harvest, it is more common for diseased plants to continue living but become stunted and produce a reduced crop of small berries.

The percentage of plants affected in any individual field usually increases significantly the year following the first appearance of symptoms. Diagnosis is made by digging up declining plants and examining their root systems, about the time that fruit begin to colour.

Abundant fleshy white roots and fine lateral roots will be seen on healthy plants, and the interior of the older woody roots is yellowish - white. With black root rot, there is usually a loss of many fine lateral roots, and irregular black patches occur along the length of the fleshy white roots. In severely affected plants, these black patches grow together so that no white roots are visible. The interior of infected older woody roots turns black.

Potential Causes

Several different fungi have been implicated as causes of black root rot, as have certain environmental stresses such as cold injury, soil compaction, and excessive water in the root zone. In some soils, black root rot has been associated with an interaction between a particular soilborne fungus and the lesion nematode Pratylenchus penetrans. It is likely that black root rot symptoms result from one or more of the following:

- Gradual buildup in the soil of disease– causing microorganisms and nematodes when strawberries are grown with inadequate rotation;
- Interaction of these organisms with environmental or other stress factors such as herbicide injury, winter or cold injury, and excessive soil moisture that might make plants more susceptible to attack; and
- Certain soil conditions such as heavy (clay) or poorly drained soils that might favour the activity of disease - causing fungi and/or inhibit the ability of the strawberry plant to produce new roots to compensate for their damage. Additional factors may also be involved.

Control

Because several factors appear to be involved in the black root - rot complex, no general control measure is totally effective. The following may help to reduce its incidence:

- Always start plantings with healthy white-rooted plants from a reputable nursery.
- Rotate out of strawberries for at least 2-3 years before replanting.
- Minimize soil compaction and increase tilth by incorporating organic matter, such as straw from a rotational grain crop.
- Avoid heavy, wet soils and improve drainage in marginal soils by tiling or planting on raised beds.
- Preplant fumigation of the soil is sometimes helpful, but not always.

PHYTOPHTHORA ROOT AND CROWN ROT OF FRUIT TREES

Phytophthora root and crown rots (sometimes called collar rot) are common and destructive diseases of fruit trees throughout the world. Apple, cherry, and peach trees are usually attacked. Pear and plum trees appear to be relatively resistant. Trees declining and dying from Phytophthora root and crown rots are frequently misdiagnosed as suffering from "wet feet" (root asphyxiation) or are sometimes confused with those suffering from winter injury.

SYMPTOMS

Diseased trees are commonly found in poorly drained areas of the orchard or yard. Heavy, wet soils that remain saturated for extended periods of time are required for disease development. Above-ground symptoms vary between tree species, but generally include reduced tree vigour and growth, yellowing or chlorosis of leaves, and eventual collapse or death of the tree. Infected trees may decline slowly over one or more years, or they may collapse and die rapidly after resuming growth in the spring. Trees may also appear healthy in the spring, but die suddenly in the latter part of the growing season.

Rapid death of trees usually occurs following excessively wet periods. On trees that decline gradually, a reddish or purple discolouration of the leaves often occurs in autumn, while leaves on healthy trees remain green. To observe below-ground symptoms, we need to remove several inches of soil around the base of the declining tree. A diagnostic reddish-brown discolouration of the inner bark and wood can be observed after cutting away the outer bark layer. A sharp line demarcates the reddish-brown (diseased) and white (healthy) portion of the crown.

Similar symptoms can be found on roots, but it is generally difficult to see root symptoms without removing the tree. This reddish discolouration and line of demarcation between diseased and healthy tissue distinguishes Phytophthora root and crown rot from other causes of tree decline and collapse such as "wet feet" (drowning) or winter injury. Roots on trees killed by excessive water are usually completely black (have no line of demarcation) and often times have an unpleasant smell. Discolouration from winter injury is usually confined to the above-ground part of the trunk, particularly on the southwest side of the tree, while the below-ground portion of the tree may still appear healthy.

Causal Organism and Disease Cycle

Phytophthora root and crown rots are caused by several Phytophthora species. These are all soilborne fungi, many of which are common inhabitants of most orchard soils. Some species that are not common inhabitants may be introduced to the orchard on contaminated planting stock or through movement of contaminated soil. While some species are much more destructive than others, depending on the type of fruit tree and rootstock, all species require extremely wet or saturated soils in order to infect and cause significant damage.

These fungi overwinter and persist in soil as mycelium in infected wood or as thick-walled spores (oospores).

Oospores remain viable in the soil for long periods of time (years). When soils are wet, oospores germinate to form thread-like fungal filaments (mycelia). Mycelia from germinated oospores or from infected tissues produce reproductive structures called sporangia. These sporangia are filled with infective spores called zoospores. Zoospores are released from sporangia only when soil is completely saturated with water (standing water).

The zoospores use flagella to swim to susceptible plant tissue where they infect. They may also swim to the soil surface and move over longer distances in runoff water. The longer the period or periods of soil saturation, the greater the risk of infection. Some rootstocks appear to be most susceptible during spring and autumn, which are also the times of year when soil temperatures are most conducive to fungus growth and zoospore production. Rootstock susceptibility and fungus activity are both low in the winter when trees are dormant.

Control

Control of Phytophthora crown and root rots is most successful using an integrated programme of cultural practices, choosing the most resistant tree species or rootstock, and when necessary, chemical control.

- Avoid sites that drain slowly or poorly or are subject to periodic flooding. Marginal sites should be modified (install drain tiles, create diversion ditches, rip underlying pan layers) to provide the additional drainage recommended for growing tree fruit crops. Planting trees on ridges or berms will raise their crowns above the primary zone of zoospore activity and provide an important margin of safety, especially in a wet year.

- Select rootstocks or tree species that are less susceptible to Phytophthora and are best adapted to our individual site. Pears are the most resistant tree fruit crop and are most likely to remain healthy in a relatively wet site. Among apple rootstocks, seedlings are relatively resistant. Among dwarfing-apple rootstocks, M-9, M-2, and M-4 are relatively resistant. The Canadian rootstock Ottawa-3 has M-9 type resistance. M-7 and MM-111 are moderately susceptible; M-26 and MM-106 are susceptible; and MM-104 is highly susceptible. Among stone fruits, plums are relatively resistant, whereas the remainder is susceptible to very susceptible. Mahaleb is the most susceptible cherry rootstock, whereas Mazzard, Morello, and Colt are somewhat more resistant and would be recommended on heavier soils
- Soil fumigation is generally considered ineffective because it never completely eradicates the fungus from orchard soils, and the Phytophthora fungi are easily reintroduced into fumigated soil.
- New fungicides have recently been developed, which are effective in controlling these diseases when used preventively, but they are seldom effective in reviving trees once the crown has become infected and moderate symptoms of decline have appeared. Fungicides are most effective when used in combination with the cultural practices described above.

BLACK ROT AND FROGEYE LEAF SPOT OF APPLE

Black rot and frogeye leaf spot are phases of a widespread and damaging disease of apple and crabapple. The fruit rot phase is called black rot and on the leaf it is called frogeye leaf spot. The disease can result in losses from:

- A rotting of fruit before harvest and in storage.
- A weakening of the tree from defoliation, and
- A blighting and dieback of twigs and limbs caused by girdling cankers.

The premature dropping of infected leaves can result in small, poor-quality fruit and reduces crop yield the following year. All apple varieties appear to be equally susceptible to fruit rot. Jonathan and Winesap appear to have the greatest susceptibility to leaf infection.

SYMPTOMS

Fruit

The disease usually starts at the calyx end of the fruit. The fungus usually enters the fruit through wounds caused by insects, hail, growth cracks, or an open calyx tube. At first, a light brown spot forms on the fruit. Usually only one spot occurs per fruit. With time, the spots enlarge and commonly develop a series of brown and black concentric bands or rings. The rotted fruit finally turns black. The decayed tissue remains firm to leathery, and holds its original shape until the entire fruit is rotted.

The completely decayed fruit finally dries and shrivels into a wrinkled black "mummy" which may remain on the tree a year or longer. Black, pimple-like fruiting bodies (pycnidia) of the causal fungus appear on the surface of rotted fruit. In cold storage, the flesh of black rot-infected fruits remains firm, in contrast to several other apple rots.

Leaves

Starting at petal-fall or somewhat later, small, purple specks appear on infected leaves. These specks enlarge to form spots 1/8 to 1/4 inch in diameter. The rounds to irregularly lobed spots develop a light brown-to-gray centre surrounded by one or more dark-brown concentric rings and a purple margin giving it a "frogeye" appearance. Black pycnidia, like those that appear on rotted fruit, may develop on the upper surface in the centres of the older leaf spots. These pycnidia help to distinguish frogeye leaf spots from similar spots caused by spray injury.

Twigs, Limbs, and Trunks

Small, slightly sunken, reddish-brown areas develop in the bark. These areas slowly enlarge and darken to form cankers. Cankers may continue to expand a little each year, and may extend down the limb for 3 feet or more. These areas remain somewhat sunken, except for the slightly raised and lobed margin.

Cankers may appear as a superficial roughening of the bark; or the bark may be killed and conspicuously cracked, especially at the margins. In recently killed areas, the bark is firmly attached to the wood; but after a year or so, it cracks and falls away and can be easily removed from the wood. Black pimple-like pycnidia and another very similar fungal fruiting structure (perithecium) are usually abundant in older cankers.

Causal Organism and Disease Cycle

Black rot and frogeye leaf spot are caused by the fungus, Physalospora obtusa. The fungus overwinters in cankers, mummified fruits, and the bark of dead wood. In the spring, the black fungal fruiting bodies (pycnidia and perithecia) release conidia and ascospores, respectively. These two types of spores spread the disease to healthy leaves, fruit, and wood. The heaviest discharge of spores occurs around blossom time, but the production of conidia may continue during wet periods throughout the summer.

The conidia can remain viable for at least one year. Leaf infection usually occurs during the petal-fall period. Conidia become attached to the leaf and may germinate in a film of moisture within 5 or 6 hours.

After germination, the fungus penetrates the leaf through natural openings in the under surface or through insect, hail or other wounds. Spore germination and infection are most rapid at 75 to 80 degrees F. Fruit infection can occur as early as petal fall; however, symptoms are usually not visible until mid to late-summer as the apple approaches maturity.

Control

Control of black rot is best achieved through an integrated programme of cultural practices and chemical control measures. Sanitation is critical for effective control. Piles of prunings are an important source of inoculum and should be removed from the perimeter of the orchard or burned. Prunings can be left on the orchard floor if they are chopped with a flail mower, which removes much of the bark and allows them to decompose faster. Removal of mummified apples and pruning out dead wood in the tree are important for reducing the inoculum within the tree.

Pruning out current-season shoots infected with fire blight is also important, because they can be colonized and serve as an inoculum source during the same growing season. Any practice that helps to maintain trees in a healthy vigorous condition is critical for controlling the canker phase of the disease.

Cankers generally develop only on stressed or weakened trees. Prune trees annually and maintain a balanced fertility programme based on soil and foliar nutrient analysis.

Cankers generally develop rapidly on winter-injured trees. The use of fungicides combined with good sanitation is beneficial for controlling the fruit rot phase of the disease. Fungicides are not effective for controlling the canker phase of the disease on weakened trees.

PHOMOPSIS CANE AND LEAF SPOT OF GRAPE

For many years, the Eastern grape industry recognized a disease called "dead-arm," which was thought to be caused by the fungus Phomopsis viticola. In 1976, researchers demonstrated that the dead-arm disease was actually two different diseases that often occur simultaneously. Phomopsis cane and leaf spot (caused by the fungus Phomopsis viticola) is the new name for the cane- and leaf-spotting phase of what was once known as dead-arm.

Eutypa dieback (caused by the fungus Eutypa armeniacae) is the new name for the canker- and shoot-dieback phase of what was once known as dead-arm.

We now propose that the name dead-arm be dropped. Growers should remember that Phomopsis cane and leaf spot and Eutypa dieback are distinctly different diseases, and their control recommendations vary greatly.

Disease incidence of Phomopsis cane and leaf spot appears to be increasing in many vineyards throughout the Midwest. Crop losses up to 30 per cent have been reported in some vineyards in growing seasons with weather conducive to disease development.

Phomopsis cane and leaf spot can affect most parts of the grapevine, including canes, leaves, and rachises (cluster stems), and flowers, tendrils, and berries and can cause vineyard losses by:

- Weakening a cane, which makes them more susceptible to winter injury?
- Damaging leaves, which reduces photosynthesis.
- Infecting cluster stems, which can result in poor fruit development and premature fruit drop.
- Infecting berries resulting in a fruit not near harvest.

Spots or lesions on shoots and leaves are common symptoms of the disease. Small, black spots on the internodes at the base of developing shoots are probably the most common disease symptom. These spots are usually found on the first three to four basal internodes. The spots may develop into elliptical lesions that may grow together to form irregular, black, crusty areas. Under severe conditions, shoots may split and form longitudinal cracks.

Although cane lesions often appear to result in little damage to the vines, it is important to remember that these lesions are the primary source of over-wintering inoculum for the next growing season. Leaf infections first appear as small, light-green spots with irregular, occasionally star-shaped, margins.

Usually only the lower one to four leaves on a shoot are affected. In time, the spots become larger, turn black, and have a yellow margin. Leaves become distorted and die if large numbers of lesions develop. Infections of leaf petioles may cause leaves to turn yellow and fall off. All parts of the grape cluster (berries and rachises or cluster stems) are susceptible to infection throughout the growing season; however, most infections appear to occur early in the growing season.

Lesions developing on the first one or two cluster stems (rachises) on a shoot may result in premature withering of the cluster stem. Infected clusters that survive until harvest often produce infected or poor-quality fruit. If not controlled early in the growing season, berry infection can result in serious yield loss under the proper environmental conditions. Berry infections first appear close to harvest as infected berries develop a light-brown colour. Black, spore-producing structures of the fungus (pycnidia) then break through the berry skin, and the berry soon shrivels.

At this advanced stage, Phomopsis cane and leaf spot can be easily mistaken for black rot. Growers should remember that the black rot fungus only infects green berries and will not infect berries after they start to mature. Berries become resistant to black rot infection by three to four weeks after bloom.

Fruit rot symptoms caused by Phomopsis generally do not appear until close to harvest on mature fruit. Severe fruit rot has been observed in several vineyards. Research has shown that berry infection can occur throughout the growing season; however, most fruit rot infections probably occur early in the season (pre-bloom to two to four weeks after bloom).

Once inside green tissues of the berry, the fungus becomes inactive (latent), and the disease does not continue to develop. Infected berries remain without symptoms until late in the season when the fruit matures. Thus, fruit rot that develops at harvest may be due to infections that occurred during bloom.

Casual Organism and Disease Cycle

The fungus overwinters in lesions or spots on old canes infected during previous seasons and require cool, wet weather for spore release and infection. The fungus produces flask-shaped fruiting bodies called pycnidia in the old diseased wood. These pycnidia release spores in early spring and are spread by splashing rain droplets to developing shoots, leaves, and clusters. In the presence of free water, the spores germinate and cause infection. Shoot infection is most likely during the period from bud break until shoots are six to eight inches long.

The optimum temperature for leaf and cane infections is between 60 and 68°F, and wetness duration of at least six hours is required at these temperatures. As the wetness duration increases, the opportunity for infection greatly increases. Lesions on leaves appear at seven to 10 days after infection. Fully expanded leaves become resistant to infection. Lesions on canes require three to four weeks to develop. The fungus does not appear to be active during the warm summer months, but it can become active during cool, wet weather later in the growing season. Pycnidia eventually develop in cane lesions and will provide the initial inoculum for infections during the next growing season.

Control

- Select planting sites with direct, all-day sunlight (avoid shade). Good soil drainage and air circulation are also very important. Orient rows to take full advantage of sunlight and wind movement. Cultural practices that increase air circulation and light penetration in the vineyard will reduce wetting periods and should be beneficial for control.
- 2While dormant pruning, cut out infected canes and destroy them. Select only strong, healthy canes that are uniform in colour to produce the next season's crop.
- Proper timing of early-season fungicide sprays is important for control in commercial vineyards.

BITTER ROT OF GRAPE

Bitter rot has been recognized as a disease of grapes since 1887. The name was derived from the bitter taste that develops in infected berries.

If 10 per cent of the berries used to make wine are infected with bitter rot, the wine may be undrinkable. The disease is more common in the southern grape growing regions of the United States, and could become a serious problem in southern region.

Bitter rot is easily confused with black rot, which is a common disease of grapes throughout. The ability to distinguish between these two diseases could be vital for growers and especially winemakers.

SYMPTOMS

Bitter rot can affect young shoots, stems of fruit clusters, pedicels and fruit. If cluster stems are infected and killed early in the season, berries do not develop and remain attached to the stem. When cluster stems are infected late in the season, berries may fall off the bunch.

This "shelling" of berries may be an important sign that bitter rot is occurring. The most obvious symptoms of bitter rot are on berries. The first symptom is a brownish, water-soaked lesion that develops on maturing fruit. Bitter rot is a ripe rot of grape, and unlike black rot, does not infect green berries.

This lesion spreads rapidly (often in concentric rings) and within a short period of time the entire fruit is affected. These infected berries usually maintain their shape and are dull brown in colour. In 2 to 3 days the skin of the berry is ruptured (more or less uniformly) by black pustules (fungal structures). Under moist conditions these pustules may coalesce to form irregular blisters over the fruit surface. The fruit skin is often torn by this point and the fruit shrivels in to black mummies closely resembling those of black rot.

If berries become infected when they are overripe, the entire berry may become infected and fall off before the pustules form.

These infected but temporarily symptomless berries are usually the most bitter, and unfortunately, the most likely to be harvested and pressed (along with healthy grapes) into wine. It is a common practice to leave wine grapes on the vine past maturity in order to increase sugar content. This could become a dangerous practice if bitter rot becomes established in the vineyard.

Causal Organism

Bitter rot of grapes is caused by the fungus Melanconium fuligineum. The fungus overwinters in the vineyard on almost any plant debris, especially berry mummies. Spores (conidia) are produced from cushions of fungus tissue (acervuli) growing on plant litter. Fungus growth and spore production are favoured by warm, humid and rainy weather. Spores are spread by splashing rain. Symptoms develop on berries one week after spores reach a healthy berry and in less time if the berry is wounded.

The fungus colonizes the entire berry, including the seeds. Acervuli develop in the fruit skin, rupturing it as they mature. Conidia in the acervuli are spread to other plant parts by rain. Bitter rot increases in severity the farther south grapes are grown.

Control

Promote good air circulation and light penetration by controlling weeds and suckers, proper pruning, and positioning or removing shoots for uniform leaf development. Where possible, rows should be planted in the direction of the prevailing wind. Good air circulation and light penetration will promote faster drying of plant parts and reduce disease incidence. Prevent wounding of berries by controlling insects, birds and other grape diseases. It is important that growers and winemakers be able to distinguish between bitter rot and black rot.

Because black rot infects only green berries, the fungicide sprays for controlling it are generally stopped when berries begin to change colour. Stopping sprays at this time could be disastrous if bitter rot is present because it infects fruit only after the berries have changed colour.

The fungicide spray programme that is recommended for control of the more common grape diseases should be beneficial in controlling bitter rot. Specific sprays for controlling bitter rot are generally not required.

CYTOSPORA CANKER OF SPRUCE

The most important disease of spruce in landscape plantings is Cytospora canker. This disease is caused by a fungus and is frequently found on Norway spruce and Colorado blue spruce and its cultivars. White spruce is also susceptible and there are a few reports on Serbian spruce. In addition to the spruces, Cytospora canker is sometimes found associated with Douglas fir, hemlocks, larches, and balsam fir.

SYMPTOMS

Dying of a lower branch with subsequent needle browning is usually the first symptom. The brown needles may remain on the branches or they may fall off. As the disease progresses over several years, higher branches show damage. The actual cankers are often first seen at the base of branches near the main trunk of the tree. On the more susceptible species (Norway spruce), trunk cankers develop which may result in girdling and death of the tree. The bark of the cankered areas is not visibly different in colour, nor does it become sunken as in cankers on many deciduous trees.

However, resin flow is usually associated with Cytospora canker and the white patches of dried resin are quite conspicuous on the bark. Resin flow can, however, be associated with any injury to branch tissue. Cankers often cannot be located without cutting into the bark. Removal of the outer bark from cankers reveals brown, dead areas of the inner bark and cambium. Within the cankered areas, a careful search using a magnifying hand lens will often reveal black pinhead sized structures that produce the spores of the pathogen.

Careful removal of layers of bark makes these structures even more visible. During wet weather, yellow to orange coloured masses of spores oozing out of these black structures in tiny tendrils can be observed. Sometimes these tendrils or gelatinous spore masses are visible to the unaided eye.

CAUSAL FUNGUS

Cytospora canker of spruce is caused by the fungus, Cytospora kunzei var. piceae (also reported as Valsa kunzei var. piceae). The spores (conidia) described above are readily disseminated by splashing water, wind-driven rain, by man during pruning, and also very likely by insects and birds. The fungus generally becomes established through wounds.

Symptoms

Cytospora canker is more common on trees over 15 years old. This disease is more prevalent on trees of low vigour. Those trees with shallow roots, weakened by drought, low fertility, mechanical injury, or insect damage; and trees growing in an unfavourable site are more susceptible to Cytospora canker. The following practices lessen the likelihood of this disease:

- Avoid bark and stem injuries.
- Control insects and mites; especially spruce gall adelgids and spider mites.
- Fertilize according to horticulturists' recommendations.
- Water during extended dry periods.
- Follow accepted pruning practices.
- Vertically mulch to relieve soil compaction, poor aeration, and inadequate water penetration.

Once established, the following may aid in suppressing disease development. Remember that affected branches cannot be saved. Prune and remove or destroy affected branches. To lessen the spread of the fungus, prune only when the trees are dry. Pruning tools should be disinfested with 70 per cent alcohol between cuts. It will generally be necessary to prune back to the main trunk. No effective chemical control measures are available.

YELLOWING, DIEBACK AND DEATH OF NARROW-LEAFED EVERGREENS

Each year various narrow-leafed evergreens such as pines, Taxus and spruce are affected with needle yellowing and browning, dieback, poor vigour or death. These are problems often associated with one or more environmental stress factors. An explanation of some of these stress factors follows.

WET SOIL

Excessive amounts of water can result in a saturated soil, reducing oxygen levels to a point where small roots weaken or die. This root decline can be sudden or gradual and the roots may be invaded by various soil-borne fungi.

Continuous wet conditions lead to progressively worse situations. If the top of the plant is unable to obtain the necessary water and nutrients, it declines or dies. However, the evidence of death (needle browning) often occurs at a much later date. In fall, winter and spring, water accumulates and literally drowns the roots. The tops of the plant may not succumb until the following spring or summer, when hot weather first arrives and stresses the plant. Some evergreens appear to lose vigour and die back after 15 to 20 years.

This often is the result of injury to the root system from moisture stresses. Heavy soils may limit development of the root system; root damage easily upsets the top-to-root ratio. Dieback and poor growth are often evident. Changes in sub-soil drainage caused by construction will often cause roots on older plants to die back.

DROUGHT OR DRY SOIL

The lack of water for long periods may result in symptoms similar to those caused by excess water. Clay soils often pull away from the roots as they dry, drying or breaking the fine roots. Drought stress may be especially noticeable in the summer on evergreens planted on well drained sites (sand or gravel), or where roots are in the top layers of heavy compacted soil. Excessive needle drop and poor vigour are often evident as a result of drought stress.

WINTER DAMAGE

Evergreen plants transpire or lose water from leaves during winter. If the soil moisture is low or roots are unhealthy, moisture in the needles is not replenished and needles are killed. Again, visible symptoms often do not appear until spring or early summer. Damage often appears on one side or on one branch of the plant, usually the side facing prevailing winds. Needles may turn brown one half or one third of the way from the tips. The extent of browning will be similar on all the needles on the branch.

DE-ICING SALT DAMAGE

Injury from exposure to de-icing salt can occur on plants. Salt sprayed by traffic on wet roads can cause browned foliage, usually on the side nearest the road. Salt solution runoff also can injure plant roots. Entire plants may die. Needle yellowing and browning often begins at the tips and gets progressively worse. Sometimes soil tests conducted in late winter indicate high salts and confirm the diagnosis.

HERBICIDE DAMAGE

Injury to evergreens by herbicides is difficult to assess. Symptoms are not always pronounced. Needle distortion may be slight, but root damage could be enough to limit water uptake. Tip damage to new growth is a common symptom with some herbicides. Look for needle distortion and twisting, or needle yellowing or browning, depending on the type of herbicide. On spruce, needle purpling and drop is common.

AIR POLLUTION DAMAGE

When atmospheric conditions allow buildup of smog or air pollutants, narrow yellowed bands may develop on the needles of susceptible plants. In other cases, the tips of the needles may turn brown. Ozone injury to white pine will cause a severe reduction in growth. If this continues for years, it is considered chlorotic dwarf disease. Trees under drought stress may be more prone to damage by air pollutants.

LOW LIGHT NEEDLE DROP

Most narrow-leafed evergreens need full sunlight. Low light conditions may result in a slow decline of some evergreens such as junipers or arborvitae. An early symptom is foliage drop in the centre of the plant. The condition is common on plants existing in overgrown, old landscapes. Sometimes two plants will grow together. Both will begin to decline. In other cases, a deciduous plant nearby may begin casting a shadow on the evergreen plant during the morning or evening hours.

TRANSPLANT OR ESTABLISHMENT PROBLEMS

Improper planting and poor after-transplant care may result in plant decline several years following transplanting. Common problems associated with planting and establishment include: burlap, especially synthetic burlap, left intact around the root ball; strings or wires left around the trunk; planting a containerized plant without disturbing the root mass; inadequate or inappropriate watering following transplanting; support wires left on the tree too long; setting the tree or shrub deeper than originally grown; and settling following transplanting. These problems are difficult to correct after symptoms have become apparent.

MANAGING ENVIRONMENTAL STRESSES

It is difficult to develop a control programme for environmental stresses. A wet soil suggests need for a better drainage system or that less watering is required. A dry soil indicates the need for a better and more uniform watering programme or using an effective mulch. Mulching, proper watering, use of a wind barrier or spraying with an antidesiccant in late fall helps lessen winter damage. Vertical mulching or core aerating will improve landscape soils.

It will hasten drainage of excessive water, preserve necessary aeration during wet periods, allow sub-soil water penetration during dry periods and promote the formation of fine feeder roots.

Drill 1 or 2-inch wide, 18-inch deep holes in the soil on 12 to 20 inch centres under affected trees. Fill holes with a mixture of equal parts of peat and a coarse aggregate, such as pumice or calcined (baked) clay particles. Control de-icing salt damage by re-directing runoff water, installing splash or spray guards or by using tolerant plants in injury prone locations.

Using Fungicide Sprays Effectively

Fungicides can be an important component of the disease management programme. However, it is important to remember that their use should be integrated with the use of sound cultural practices, knowledge of pathogen and disease biology, and disease resistance whenever possible.

Fungicides are only effective when infectious plant diseases that are caused by fungi are truly the cause of the problem. In many cases, pests and diseases follow other environmental imbalances and may not be the major problem. In cases such as these, a fungicide may help but is often not the total answer. Also, it is important to remember that fungicides are only effective if several rules are followed. First, the correct material must be selected.

This depends on correct diagnosis and identification of the pathogen. Second, the chemical must be applied at the right time of year and frequently enough to protect plant material adequately. Third, fungicides must be applied properly over plant surfaces. These three rules depend on making correct decisions based on correct knowledge. Too many people simply "spray and pray," and are often disappointed with the results.

Correct Diagnosis

One must be sure of what the problem is before proceeding. The most effective fungicides in use today have been developed for specific situations and specific diseases. To use these chemicals, we must spend time making a correct diagnosis. Our state Extension specialists, plant disease diagnostic clinic, and county Extension agents can assist we with the proper diagnosis.

Selecting the Proper Material

Diagnosis leads to selection of the right material to do the job. Usually, several materials are effective against the type of disease we are dealing with. For instance, triforine, sulfur or triadimefon all control powdery mildews. Before selecting any chemical, read the label.

Can we carry out the instructions? Is the plant type listed on the label? If so, the chemical is registered for use on the plant, and should be effective in providing disease control if used properly. If not, it is illegal to use that particular pesticide.

Use the Correct Method of Application

Foliar application fungicide sprays usually work in controlling infectious diseases because they act as a chemical barrier on leaf, stem or flower surfaces. When the pathogen arrives on the plant surface, it encounters this barrier and is unable to infect the plant. Effective fungicide use requires that this barrier be as complete as possible. A spray method must provide the best combination of practical usefulness and good coverage.

For many diseases, special attention must be given to undersurfaces of leaves, especially on the lower leaves of the plant. The completeness of the barrier depends on how well the spray spreads and sticks to the plant surfaces. For this reason, spreader-stickers or spray adjuvant can be added to many sprays.

Sometimes the product label alerts the user to these problems. However, observing the spray deposit after we have finished some of the job may be the best way to decide if an adjuvant should be used. Hairy or waxy foliage is especially difficult to cover properly without a spreader-sticker.

Proper Timing of Fungicide Applications

Timing refers to when and how often the spray must be applied to effectively control a disease. The first application usually is made at a time close to but before the pathogen arrives on the plant surface.

This information is often provided on the pesticide label or is available in Extension literature from our local county Extension agent. In most situations, fungicides are not effective in controlling the disease if the pathogen has already entered (infected) the plant tissues. In many cases, specific information about the disease cycle may be needed to time the first application correctly. After the first application is made, the pesticide barrier is established on the plant surfaces.

Effective use involves keeping this barrier active and complete throughout the time that the pathogen can arrive on and infect the plant. Modern fungicides are developed so that they do not persist in the environment for long periods of time. Rainwater, sunlight, microbial action and oxidation decrease effectiveness of the fungicide.

Reapplication of the spray is needed in many cases to keep the fungicide barrier active. Plant growth also affects the completeness of the barrier. As new leaves and shoots appear, they are unprotected and may be subject to infection. If so, they must be recovered with the barrier. The fungicide label gives reapplication guidelines, usually in ranges of 7-14 day intervals. If excessive rainfall or rapid growth of the plant occurs, the shorter interval between sprays should be used. If not, use the longer interval.

LEAF BLIGHT OF HAWTHORN

Leaf blight is a serious disease of a few very susceptible English hawthorn cultivars. Most other hawthorn species and cultivars are resistant or are not seriously affected.

SYMPTOMS

The disease is first evident as small, angular, reddish-brown spots on the leaves. The spots have irregular margins and coalescence of spots often occurs resulting in larger irregular diseased areas. Diseased leaves yellow and fall prematurely. Sometimes, spots may be more prevalent near the margins of the leaves.

When conditions are favourable for disease development, extensive defoliation occurs. It is usually mid-summer or after before the disease becomes prevalent. Small, black, raised dots develop in the centre of the spots. These are the spore masses of the causal agent. They are especially evident when the leaves are wet.

Causal Fungus

Leaf blight of hawthorn is caused by the fungus, Entomosporium thuemenii (Diplocarpon maculatum). The fungus survives from one year to the next in fallen diseased leaves and in inconspicuous stem spots. During May and June, spores are produced on these over-wintered leaves. These spores are spread by splashing rain and initiate the disease on the current season's foliage. As the leaf spots develop, new spores are formed on the spots and rapidly spread the disease. Wet weather is favourable for rapid development because splashing water carries spores, and persistent water drops favour spore germination and infection.

Control

Plant Resistant Varieties

Washington hawthorn types are resistant to this disease. English hawthorn types are susceptible and are commonly infected. Although Washington types are resistant to leaf blight they are susceptible to rust diseases. If hawthorn rust is a serious problem in our area, it may be best to select plants other than hawthorns for landscape plantings. Since the fungus over-winters in the fallen diseased leaves, raking and destroying all leaves will help to manage leaf blight, but will probably not result in complete control.

Protection with Fungicides

Leaf blight can be prevented by spraying with fungicides. Control through use of fungicides depends on proper timing of the sprays. Based on presently available information, spraying at 10 to 14 day intervals from bud break through early July has given good control. Additional applications may be necessary during rainy seasons or when good control was not achieved with the earlier sprays.

DIPLODIA TIP BLIGHT OF AUSTRIAN, RED AND SCOTCH PINE

This disease is most commonly seen on Austrian pine and

some of the other two-and three-needle pines such as red pine, Mugho pine and Scots pine. It is found more uncommonly on white pine, spruces and other evergreens. The fungus commonly attacks mature trees that have been under stress from drought, root restriction or other planting site problems. It can also be a problem in young, rapidly growing nursery or Christmas tree plantings.

SYMPTOMS

The pathogen infects and kills current year's shoots. When the infected, current season's needles are l/2 to 3/4 expanded, they turn yellow, then brown as they die on individual branch tips. A close look at the bases of the dead needles may reveal tiny black, fungal fruiting bodies emerging from the needle surface. Repeated infection over several years causes the ends of affected branches to have a proliferation of shoots.

If left unchecked the disease can eventually kill mature trees. Girdling cankers can be formed if the pathogen infects wounds on the stem. Other problems can cause similar dieback and tree decline. Winter drying; drought; injury from weevils, pine-shoot moths or tip moths; and some needlecast diseases caused by other fungi may cause damage similar to tip blight.

Causal Fungus and Disease Development

Tip blight is caused by the fungus, Sphaeropsis sapinea, once known as Diplodia pinea. Spores of the fungus develop in the black fruiting bodies located at the base of infected needles and other affected plant parts from spring through fall. They are spread about only during periods of rainfall. Pine shoots are particularly susceptible to infection in early spring. Developing cone scales are also commonly infected, although they are not damaged. Wounds, such as those made by hail, shearing, or insects (weevil or spittlebug feeding) also serve as entry points for the fungus. The fungus survives over winter in the infected shoots, bark, cones or needle litter beneath the tree.

Control

- Trees should be kept in good vigour with regular

maintenance, deep watering during droughts, fertilizing, control of insects and vertical mulching to open up the soil in the root zone. Vertical mulching can be done to improve landscape soils. Vertical mulching will lessen damage due to excessive water, preserve necessary aeration during wet periods, allow sub-soil water penetration during dry periods, and promote the formation of fine feeder roots. Drill one or two inch wide, 18" deep holes in the soil on 12-20" centres under affected trees near the drip line of the branches (where fine feeder roots are located). Fill holes with a mixture of equal parts of peat and a coarse aggregate such as pumice or calcined clay particles.

- Remove previously blighted shoots. Since many spores are produced on cones, removal of previously blighted shoots probably does not decrease spore numbers appreciably. However, it does serve to make the tree look better and may increase its vigour.
- Do not shear or prune infected trees during wet weather because spores released at this time may be carried from tree to tree on pruning tools.
- This disease can be partially controlled with fungicides. Attention must be given to protecting the new spring growth of the trees from bud swell to full candle elongation. Make first application just prior to bud break and make two or more additional applications at 10-day intervals. It is important to get the first application on the trees before any bud sheaths have broken. If bud sheaths have broken, spraying with fungicides is a waste of time and money.

MAINTAINING HEALTHY RHODODENDRONS AND AZALEAS IN THE LANDSCAPE

Rhododendrons and azaleas, which are closely related, are among the most popular flowering shrubs. In many urban landscapes, these shrubs thrive and have relatively few serious health problems once they are established. However, they do have some "special requirements" that must be met to insure good health. These special conditions match those in areas where rhododendrons and azaleas are native.

Rhododendrons and azaleas grow on forest floors in many parts of the world, in shaded habitats, with acidic soils rich in organic matter. Soils are often covered with a surface layer of decaying leaf litter. Matching these conditions where native rhododendrons and azaleas thrive is the key to their good health in the landscape. Plant in areas with good soil drainage, acid pH and partial shade that are sheltered from direct afternoon sun and winter winds. Some of the more common problems that can occur on rhododendrons and azaleas include: iron deficiency, winter injury (burn), black vine weevil, and Phytophthora root rot.

IRON DEFICIENCY

The yellowing of rhododendron leaves typical of iron deficiency. The yellowing is between the veins and more severe on younger leaves. This problem generally results from plants growing in soils of improper pH. Rhododendrons must be grown in acidic soil that is high in organic matter. If the pH is above 6.0, soil amendments such as sulfur, iron sulfate or ammonium sulfate must be incorporated into the root area to lower pH. It will be difficult to overcome the deficiency problem in soils high in lime or calcium, even with soil amendments. In such cases, mulch the plant heavily with a good grade of sphagnum peat.

Bark mulch mixed with the peat provides mulch with good aeration and drainage. It also suppresses root rotting organisms. If kept moist, plants will root into this mulch. Use a complete, acid fertilizer that contains iron.

WINTER BURN

Leaf drying and browning can occur on rhododendron leaves as a result of winter exposure. The leaves, even though they may be "rolled up" at times, are subject to drying out in dry winter air. The solution is to protect the plant from the drying wind. Plant rhododendrons behind buildings or other plants that can serve as wind shields. Put wind shields in place around the plants during the winter months. Mulching as described above is critical to preventing winter injury.

BLACK VINE WEEVIL

The adult weevils feed on rhododendron leaves producing a C-shaped notching in the leaf margin. These insects can be quite damaging. Most of the damage comes from weevil larvae feeding on the roots. Affected plants lose vigour, and may die eventually. The insecticide Orthene is registered for use on rhododendrons, and it will control black vine weevil adults.

PHYTOPHTHORA ROOT ROT (RHODODENDRON WILT)

This disease is caused by a soil-borne fungus (Phytophthora). Generally, it is a problem where wet (saturated) soil conditions occur frequently. Early symptoms of the disease consist of retarded growth, drooping of foliage (perhaps on one or two branches only) and yellowing of leaves. Infected roots appear dark and "mushy."

As the disease progresses, a browning discolouration of the wood may extend upward from the base on affected branches. Plants in poorly drained soils are very subject to water-logging which makes them highly susceptible to this disease. If this occurs, plants may die quickly. Whereas infected plants cannot be cured, root rot may be tolerated by the plant if improvements in soil drainage and aeration are made as soon as possible.

Young plants can be lifted and replanted. Before replanting, improve the drainage and aeration of the soil. Use tile drainage or add porous materials in a layer beneath the root zone. Plant in a raised bed and do not mound the soil up around the crown. Mulching with tree bark provides biological control. The mulch must be applied to a depth of two inches and reapplied as it decomposes.

Chapter 4

Effect of Seed Born Diseases on Agriculture

SEPTORIA LEAF SPOT OF TOMATOES

Septoria leaf spot is a disease of the foliage and stems. It does not affect fruit directly. The disease causes rapid defoliation when weather is warm and moist.

SYMPTOMS

Although the symptoms may appear on the leaves and stems at any stage of plant development, they usually become evident after plants have begun to set fruit in mid-July. Initially the fungus causes small, water-soaked, roughly circular spots scattered randomly over the leaf. These spots enlarge to become 1/16 to 1/8 inch in diameter with dark margins and tan centres. Small, dark, pimple-like fruiting bodies in which spores of the fungus are produced can be seen in the centre of the lesions.

Older leaves near the ground usually show symptoms first. Symptoms appear rapidly on younger foliage in rainy weather. When leaves are heavily infected, they drop prematurely. Exposed fruits are subject to sunscald. The fungus which causes this disease can grow on weeds such as Jimson weed, horse nettle and nightshade.

Causal Organism

Septoria lycopersici lives between tomato crops in the soil

on infested debris of tomato and weeds. Spores formed on crop debris splash onto foliage and start the disease. Wind and rain spread spores produced in the dark bodies formed in leaf spots to adjacent uninfected leaves. The fungus is most active between 60 and 80 degrees F when rainfall is abundant.

Control

- Rotate out of tomatoes for 4 years.
- Deep plow, preferably in the fall, to bury all plant refuse.
- Grow tomato transplants in sterilized soil.
- Control weeds, especially horse nettle, Jimson weed, and nightshade.
- Use of a protectant fungicide may be necessary to adequately control Septoria Leaf Spot when conditions are favourable for disease development.

FUSARIUM WILT OF VINE CROPS

Several different species of Fusarium, a soil-borne fungus, cause wilting of watermelon, muskmelon, cucumber, squash, and other vine crops. In many cases the fungus-causing wilt in a particular crop is specific to that crop. These fungi are generally capable of surviving for long periods in the soil.

SYMPTOMS

Wilting of older plants is often the first symptom of the disease. Before diseased plants totally collapse, however, they may begin to wilt during the hottest part of the day and recover during the night. Infected plants are often stunted and yellowed. Leaves often have dead areas which can mimic nutrient deficiencies. Stems of wilted plants when cut lengthwise at the soil line may show brown discolouration in the woody tissues immediately under the bark. Vines killed by Fusarium can be covered with pinkish-white fungal growth in wet weather.

Causal Organism

Fusarium oxysporum is well adapted to life in the soil.

It can survive season to season in old diseased vines or it can live free in infested soil for many years in the absence of its host crop. The fungus grows at soil moisture and temperature favourable for vine crop growth. If the soil is very wet, infection is reduced. The fungus is stimulated to germinate when roots of susceptible host plants are growing nearby. It enters the plant through root tips or where some opening is present.

Control

- Plant wilt-resistant cultivars whenever possible. Degree of resistance is influenced by the populations of the fungi in the soil, and which races are present.
- Avoid introduction of the fungus into fields. Once soil is infested, crop rotation will be of little use because of the long-term survival of these fungi in soil. The fungus can be spread on equipment, tools, feet, and surface water contaminated with infested soil. Do not put compost on fields which has been made from diseased vines. Compost resulting from such piles will contain the fungus.
- Crop rotations of three to four years may be helpful in lowering the amount of Fusarium in the soil. Since the types of Fusarium-causing wilt in a crop are generally specific to that crop it is possible to rotate between vine crops in infested soil.
- In commercial production of vine crops, soil fumigation may be useful.

ANTHRACNOSE OF TOMATO

Anthracnose is a common and serious disease of tomato fruit. This disease can occasionally cause severe damage to peppers, especially when red fruit is allowed to develop. Anthracnose can reduce a bountiful harvest into rotted fruit in a few days in warm, moist weather.

SYMPTOMS

Small, water-soaked, circular lesions develop under the

skin of fruit as it ripens. These become sunken and dark. Numerous dark specks, the fruiting bodies of the fungus, develop in the lesions in concentric rings. In moist, warm weather, these black bodies ooze gelatinous pink spore masses. In warm weather the fungus and soft rot bacteria which enter the split skin over the lesions spread internally forming a semisoft decay which renders the fruit worthless.

Causal Organism

Colletotrichum coccodes survives between crops on infested plant debris in the soil. Early in the growing season, spores from the soil splash on lower leaves of the tomato plant. Few symptoms develop on infected leaves, but the spores produced on foliage can be carried by splashing rain to developing green fruit. Infected green fruit will not develop symptoms of anthracnose until they begin to ripen. Ripe fruit is very susceptible to this fungus.

Control

- Rotate 3 years between pepper and tomato crops.
- Plant tomatoes and peppers in well drained fields to avoid excess soil moisture as fruit ripen.
- Apply overhead irrigation during the early part of the day so that plants dry before sundown.
- Harvest and use fruit before it fully ripens.
- If conditions favour development of anthracnose, a preventative spray programme may be required to give adequate control of this disease.

Apply registered fungicides according to product label instructions when weather conditions are above 65 degrees F and the foliage is likely to remain wet longer than 6 hours. Applications to tomatoes should begin when the first fruit is larger than a walnut. Applications to peppers should be started as soon as fruit is present.

ROOT KNOT NEMATODE

Root knot first attracted attention in 1855 when it was found in a greenhouse in England. Since then it has been a cause for concern over much of the world. It affects more than 2,000 species of plants including economically important forage crops, small grains, fruits, vegetables, field crops, nursery crops, and turf grasses. Root knot nematodes are cosmopolitan in distribution and occur widely in the United States.

Symptoms

Root knot is very distinctive because of the galls or swellings produced on roots and underground portions of stems. These deformations can often completely ruin crops for sale. Plants, if infected when young, will be stunted, more susceptible to drought stress, and show symptoms of nutrient deficiency. Large and small roots may be affected with swellings varying from round sphere-like galls to elongated spindles formed from large numbers of individual galls growing together. Root knot galls involve the entire root in the affected zone. They do not take the form of easily detached galls like those produced by nitrogen-fixing bacteria on the roots of legumes. Root knot galls should not be confused with the fungal disease, club root, on cruciferous crops. In most instances, root knot is characterized by smaller swellings, and more uniformly distributed infection on the lateral feeding roots than is typically seen with clubroot.

When the galls formed by the root knot nematode are broken open, shiny white bodies about the size of a pinhead, the enlarged female nematodes, are usually found. Also, glistening white to yellow egg masses are present on the root surface. The galls formed by clubroot do not possess this characteristic and are usually pinkish or brick coloured. Phenoxy-type herbicides such as 2,4 D can cause swellings on stems of cruciferous, crops which superficially look like root knot galls. Herbicide damage on these crops will not affect the roots, however. On young potato tubers the outer surface appears rough and warty because of the enlarged females underneath the skin.

Causal Organism

Root knot is caused by a small round worm, Meloidogyne

spp. There are about 40 species at the present time. Not all plants are susceptible to any one species.

Juveniles emerging from eggs in the soil penetrate between and through cells to a position at the centre of the root usually near the growing tip. Feeding by the nematodes causes increases in root cell numbers and size. Enlarged cells, called giant cells, serve as food sources for female root knot nematodes. During feeding, juveniles which will become females undergo a series of molts and enlarge.

The female does not move from the feeding site. Each female deposits 300-500 eggs in a protective, jelly-like matrix at the root surface. Eggs, particularly in the egg mass, can withstand unfavourable environmental conditions. They represent one means of overwintering. Root knot nematodes may also overwinter in the soil in a juvenile stage. Root knot nematodes are often introduced into a growing area in infected planting stock.

Control

- Rotate at least 3 years with non-host crops. While the feeding habits of field populations must be determined before crop rotation plans are set, the most common root knot species will not reproduce on grasses, small grains, or corn. Clean cultivation to keep down weed hosts of the nematodes is important.
- Use transplants and nursery stock certified free of root knot nematode.
- Plant resistant varieties wherever possible.
- In commercial production, soil can be treated with fumigant-type nematicides.

EARLY BLIGHT

Early blight is a very common disease of both potato and tomato. It causes leaf spots and tuber blight on potato, and leaf spots, fruit rot and stem lesions on tomato. The disease can occur over a wide range of climatic conditions and can be very destructive if left uncontrolled, often resulting in complete defoliation of plants. In contrast to the name, it rarely develops early, but usually appears on mature foliage.

SYMPTOMS

On leaves of both crops, the first symptoms usually appear on older leaves and consist of small, irregular, dark brown to black, dead spots ranging in size from a pinpoint to 1/2 inch in diameter. As the spots enlarge, concentric rings may form as a result of irregular growth patterns by the organism in the leaf tissue. This gives the lesion a characteristic "target-spot" or "bull's eye" appearance.

There is often a narrow, yellow halo around each spot and lesions are usually bordered by veins. When spots are numerous, they may grow together, causing infected leaves to turn yellow and die. Usually the oldest leaves become infected first and they dry up and drop from the plant as the disease progresses up the main stem. On tomato, stem infections can occur at any age resulting in small, dark, slightly sunken areas that enlarge to form circular or elongated spots with lighter-coloured centres.

Concentric markings, similar to those on leaves, often develop on stem lesions. If infested seed are used to start tomato transplants, seedlings may damp off soon after emergence. When large lesions develop at the ground line on stems of transplants or seedlings, the plants may become girdled, a condition known as "collar rot." Such plants may die when set in the field or, if stems are weakened, may break over early in the season.

Some plants may survive with reduced root systems if portions of stems above the canker develop roots where they contact the soil. Such plants, however, usually produce few or no fruits. Stem lesions are much less common and destructive on potato. Blossom drop and spotting of fruit stems, along with loss of young fruit, may occur when early blight attacks tomatoes in the flowering stage.

On older fruits, early blight causes dark, leathery sunken spots, usually at the point of stem attachment. These spots may enlarge to involve the entire upper portion of the fruit, often showing concentric markings like those on leaves. Affected areas may be covered with velvety black masses of spores. Fruits can also be infected in the green or ripe stage through growth cracks and other wounds. Infected fruits often drop before they reach maturity.

On potato tubers, early blight results in surface lesions that appear a little darker than adjacent healthy skin. Lesions are usually slightly sunken, circular or irregular, and vary in size up to 3/4 inch in diameter.

There is usually a well defined and sometimes slightly raised margin between healthy and diseased tissue. Internally, the tissue shows a brown to black corky, dry rot, usually not more than 1/4 to 3/8 inch deep.

Causal Organism

Early blight is caused by the fungus, Alternaria solani, which survives in infected leaf or stem tissues on or in the soil. This fungus is universally present in fields where these crops have been grown. It can also be carried on tomato seed and in potato tubers.

Spores form on infested plant debris at the soil surface or on active lesions over a fairly wide temperature range, especially under alternating wet and dry conditions. They are easily carried by air currents, windblown soil, splashing rain, and irrigation water.

Infection of susceptible leaf or stem tissues occurs in warm, humid weather with heavy dews or rain. Early blight can develop quite rapidly in mid to late season and is more severe when plants are stressed by poor nutrition, drought, or other pests. Infection of potato tubers occurs through natural openings on the skin or through injuries. Tubers may come in contact with spores during harvest and lesions may continue to develop in storage.

- Use a crop rotation that includes potatoes or tomatoes only every third or fourth year to allow infested plant debris to decompose in the soil. Rotations with small grains, corn or legumes are preferable.

- Use tillage practices such as fall plowing that bury all plant refuse.
- Select cultivars that have a lower susceptibility to early blight.
- Use certified disease-free tomato seed and transplants.
- If producing tomato transplants, disinfest soil in plant beds and control humidity in cold frames or greenhouses. Practice good sanitation throughout the transplant production operation.
- Use appropriate measures to control weeds and volunteer potatoes and tomatoes in production areas.
- Maintain fertility at optimal levels-nitrogen and phosphorus deficiency can increase susceptibility to early blight.
- Time applications of overhead irrigation to allow plants to dry before nightfall.
- Although the above measures are important to minimize infection, it is usually necessary to apply fungicide sprays to fully protect plants from early blight.

  Applications to tomato are usually begun 2-3 weeks following emergence or soon after transplanting if a calendar schedule is followed. Timing of fungicides can also be made using the TOM-CAST system. For potatoes, fungicide applications should be initiated when plants begin to flower. Thorough coverage is important. Adjust equipment for good vine penetration.

LATE BLIGHT OF POTATO AND TOMATO

Late blight is one of the most devastating diseases of potato and tomato worldwide. It was responsible for the devastating Irish potato famine of the 1840's and has continued to be important to the present. Since 1990, late blight has caused widespread damage across the United States and Canada. If left unmanaged, this disease can result in complete destruction of potato or tomato crops.

SYMPTOMS

Late blight appears on potato or tomato leaves as pale green, water-soaked spots, often beginning at leaf tips or edges. The circular or irregular leaf lesions are often surrounded by a pale yellowish-green border that merges with healthy tissue. Lesions enlarge rapidly and turn dark brown to purplish-black. During periods of high humidity and leaf wetness, a cottony, white mold growth is usually visible on lower leaf surfaces at the edges of lesions.

In dry weather, infected leaf tissues quickly dry up and the white mold growth disappears. Infected areas on stems appear brown to black and entire vines may be killed in a short time when moist weather persists.

On potato tubers, late blight appears as a shallow, coppery-brown, dry rot that spreads irregularly from the surface through the outer 1/8-1/2 inch or more of tissue. On tuber surfaces, lesions appear brown, dry, and sunken, while infected tissues immediately beneath the skin appear granular and tan to copper-brown. When tubers are stored under cool, dry conditions, lesion development is retarded and, upon prolonged storage, lesions may become slightly sunken and desiccated.

Secondary bacteria and fungi frequently enter late-blight lesions, usually resulting in a slimy breakdown of entire tubers. Late blight can also develop on green tomato fruit, resulting in large, firm, brown, leathery-appearing lesions, often concentrated on the sides or upper fruit surfaces. If conditions remain moist, abundant white mold growth will develop on the lesions and secondary soft-rot bacteria may follow, resulting in a slimy, wet rot of the entire fruit.

Causal Organism

Late blight is caused by the fungus Phytophthora infestans. Unlike most pathogenic fungi, the late blight fungus cannot survive in soil or dead plant debris. For an epidemic to begin in any one area, the fungus must survive the winter in potato tubers (culls, volunteers), be reintroduced on seed potatoes or tomato transplants, or live spores must blow in with rainstorms. Disease development is favoured by cool, moist weather. Nights in the 50's and days in the 70's accompanied by rain, fog or heavy dew are ideal.

Under these conditions, lesions may appear on leaves within 3-5 days of infection, followed by the white mold growth soon thereafter. Spores formed on the mold are spread readily by irrigation, rain and equipment. They are easily dislodged by wind and rain and can be blown into neighbouring fields within 5-10 miles or more, thus beginning another cycle of disease. Infection of potato tubers arises from spores that develop on foliage. Tubers exposed by soil cracking or erosion of hills may come in contact with spores washed down from infected leaves and stems by rainfall or irrigation.

Tubers infected during the growing season may partially decay before harvest. Tuber infection may also occur at harvest when tubers contact living spores remaining on infected vines. Little if any tuber-to-tuber spread of late blight occurs during storage if tubers are kept under cool, well-ventilated conditions. The protectant fungicides commonly used to protect plants from late blight remain fully effective with all known strains of the fungus.

- Infected cull potatoes are a major source of spores of the late blight fungus and must be disposed of properly-Do Not Make Cull Piles. Cull potatoes should be spread on fields not intended for potato production the following year in time that they will totally freeze and be destroyed during the winter. If this is not possible, they must be destroyed in some other way such as by complete chopping, burial, burning or feeding to livestock.
- Plant only certified seed potatoes. Use of "year-out" seed or seed saved from local crops is asking for trouble with late blight. Seed sources should be selected very carefully to avoid bringing in late blight on seed potatoes, especially new strains of the fungus. Look for the characteristic coppery-brown discolouration of the potato flesh under the skin of seed tubers.

  Infected tomato transplants also can be a significant

source of the disease. Use only obviously healthy tomato transplants free of dark lesions on leaves or stems.

- Volunteer potatoes and tomatoes can be a significant source of spores of the late blight fungus. All volunteers should be destroyed as quickly as possible by herbicides, chopping, or cultivation.
- Growers should scout fields regularly to look for late blight. Special attention should be paid to early-planted fields because that is where the disease is likely to develop first. Scouting should be concentrated in low-lying areas, field edges along creeks or ponds, near the centre of centre-pivot irrigation rigs, in areas near woodlots or any area that is protected from wind where the leaves tend to remain wet longer.

  Any area where it is difficult to apply fungicides such as edges and corners or under power lines if using aerial application should be examined. Scouts should look for large, black or purplish lesions on stems or leaves and the telltale cottony, white mold growth, usually on the undersides. Be sure to check leaves and stems under the crop canopy as that is where the disease is most likely to begin.
- Use of a good protectant fungicide programme is necessary to fully protect any crop of potatoes or tomatoes.
- With potatoes, make sure that vines have been completely dead for 2-3 weeks prior to harvest. Fungicide applications should be continued until vines are dead.

  When foliage dies, spores of the late blight fungus that remain on the foliage also die. This practice will prevent infection of tubers during harvest and development of late blight in storage.

BACTERIAL RING ROT OF POTATOES

Bacterial ring rot is an important disease of potatoes and

is one of the main reasons for rejection of seed potatoes from certification programmes. This disease is particularly serious because it has the potential to spread quickly throughout a farm and may lead to severe losses if left unchecked. Ring rot was originally found in Germany in the late 1800's. The causal bacteria were introduced into the United States in the early 1930's and by 1940 were found throughout the country.

SYMPTOMS

Severe ring rot can result in wilting of leaves and stems along with yellowing and death of leaves. Lower leaves usually wilt first, are slightly rolled at the margins, and are paler green than healthy leaves. As wilting progresses, leaf tissues between veins become yellow. In the later stages of disease, margins of lower leaves die and become brittle and eventually entire stems yellow and die.

Frequently, only one or two stems in a hill will develop symptoms and, in some cases, there are no above-ground symptoms at all. Ring rot derives its name from a characteristic breakdown of the vascular ring within the tuber. This often appears as a creamy-yellow to light-brown, cheesy rot. The symptom is most frequently observed when a diseased tuber is cut crosswise at the stem end. In severe cases, the vascular ring may be separated, and a creamy or cheesy exudate can be forced out from this tissue when the tuber is squeezed.

On the outer surface, severely diseased tubers may show slightly sunken, dry, cracked areas. Infected tubers are often invaded by secondary decay organisms which may lead to complete breakdown. Symptoms of ring rot in the vascular tissue of infected tubers are often less obvious t, appearing as only a broken, sporadically appearing dark line, or as a continuous, yellowish dis-colouration. Because of this, laboratory tests should always be performed to confirm a diagnosis of ring rot.

Causal Organism

Ring rot is caused by the bacterium Clavibacter michiganense subsp. sepedonicus. Ring-rot bacteria survive between seasons mainly in infected seed tubers. They are also capable of surviving 2-5 years in dried slime on surfaces of crates, bins, burlap sacks, or harvesting and grading machinery, even if exposed to temperatures well below freezing. Survival is longest under cool, dry conditions. Ring-rot bacteria do not survive in soil in the absence of potato debris, but can survive from season to season in volunteer potato plants.

Wounds are necessary for penetration of the bacteria into seed pieces. The pathogen is easily transmitted from diseased tubers to healthy seed pieces during the seed-cutting process. A knife that cuts one infected tuber can spread these bacteria to the next 20-100 seed pieces. Likewise, the bacteria may be spread during planting, particularly if a picker-type planter is used. Ring-rot bacteria can be moved in irrigation water and by chewing insects, such as Colorado potato beetles and flea beetles.

After the bacteria become established in a plant, they multiply and move throughout the plant via the water-conducting tissues. Fortunately ring-rot bacteria are capable of causing disease only in potato, although they may be able to colonize roots of sugar beets.

- Plant only certified disease-free seed tubers. Certified seed potatoes are produced under regulations mandating zero tolerance for ring rot. Although use of certified seed tubers will not guarantee total freedom from ring rot bacteria, it is the best assurance.
- Discontinue use of any lot of seed tubers in which ring rot is found. Seed lots known to be contaminated with ring-rot bacteria should never be planted.
- Before handling seed tubers, all containers, tools, knives and mechanical cutters, planters, and other equipment should be thoroughly washed with a detergent solution, rinsed, and then sanitized with a disinfectant (for current recommendations,
- If ring rot is confirmed to be present, a thorough cleanup must be undertaken. Dispose of all infected tubers away from potato production areas. Clean all

surfaces of storages and equipment to remove all mud, dirt and debris and then wash with a strong detergent in hot water applied by a high-pressure washer. After cleaning, sanitize all storages and equipment with a disinfectant. Do not plant potatoes for two seasons in any field in which ring rot has been found.

POTATO PINK ROT, PYTHIUM LEAK AND SEED-PIECE DECAY

Pink rot and Pythium leak, sometimes collectively called water rot; occur sporadically wherever potatoes are grown. These diseases are a problem of mature tubers at harvest and in storage. They are most serious when warm, wet soil conditions persist during tuber formation and at harvest. When newly-planted seed pieces are exposed to these conditions, Pythium seed-piece decay also can be severe. Major problems with these diseases are usually associated with excessive rainfall or irrigation either early or late in the season, especially on poorly-drained soils.

SYMPTOMS

Pythium seed-piece decay often results in delayed emergence and poor stands. Infected seed pieces become a soft, watery mass in the soil. Symptoms of pink rot in mature plants include brown or blackened roots or stolons, and in severe cases, leaf chlorosis, stunting, wilting, and even plant death. Tubers develop pink rot mostly through diseased stolons, but occasionally infections occur at buds or lenticels. Decay spreads through infected tubers with the advancing margin of the rot usually sharply defined by a dark line, which may be visible through the skin. Eyes of infected tubers are often dark brown.

Decaying tubers remain intact, but are spongy and odorless. If squeezed, a clear liquid will exude. When infected tubers are cut open, the internal tissues turn salmon pink after a 15-20 minute exposure to air, then later become brownish-black.

Pythium leak usually develops through harvest wounds in tuber surfaces and begins as a discoloured, watersoaked area. As with pink rot, the advancing margin of infection is usually bounded by a dark line. Infected tissues are extremely watery, and appear brown or gray. Severely rotted tubers are of a uniform texture resembling a soft, watery paste.

Causal Organisms

Pink rot is caused by several species of the soilborne fungus Phytophthora while Pythium leak and seed-piece decay are caused by several species of the closely related soilborne fungus Pythium. These fungi are widely distributed in both water and soil, and their behaviour is similar. They survive in soil within decaying plant material, or as resistant spores free in the soil. In warm, moist soil, these fungi produce swimming spores that move in water films.

Roots can be infected by Phytophthora at almost any stage of plant growth, but symptoms are more severe on younger roots. Both groups of fungi infect tubers through wounds, but Phytophthora generally infects tubers before harvest, often through stolons. Infection by Pythium usually occurs through harvest wounds, especially at temperatures above 70 degree F. Seed pieces can be infected by Pythium as soon as they are planted.

- Select areas with well-drained soils for planting potatoes.
- Use a crop rotation away from potatoes for at least 4 years if pink rot or leak has been severe. This may reduce the amount of fungus surviving in the soil.
- Delay planting for at least 2 weeks after plowing down green vegetation as this may temporarily stimulate populations of Pythium fungi.
- Avoid planting in soils colder than 45 degree F or warmer than 70 degree F.
- Avoid harvesting infested fields when soils are especially wet or soil temperatures are below 50 degree F or above 65 degree F. Stop irrigation well in advance of harvest.

- Avoid bruising tubers during harvest by adjusting equipment properly, keeping digger chains fully loaded and minimizing drops to 6 inches or less. Do not leave harvested tubers lying on warm, moist soils for any length of time as infection with Pythium may occur quickly.
- Leave low spots in fields unharvested if they have been waterlogged and much rot is present.
- Keep tubers cool and as dry as possible during harvest, loading, transit and storage.
- Grade out infected tubers as much as possible before placing harvested tubers in storage.
- Store lots of harvested tubers containing many infected tubers separately from healthy lots. Good airflow through the pile should be provided to dry out leaky tubers. Lots with significant amounts of disease should be marketed as soon as possible as they will not store well.

SCAB OF POTATO TUBERS

Scab is a disease of potato tubers that results in lowered tuber quality due to scab-like surface lesions. There are no above-ground symptoms. Two forms of scab occur. Common scab occurs in all production areas and is most severe in soils with a pH above 5.5. Another less common form, called acid scab, is important in acidic soils.

SYMPTOMS

Scab symptoms are quite variable. Usually, roughly circular, raised, tan to brown, corky lesions of varying size develop randomly across tuber surfaces. Sometimes scab develops as a rather superficial layer of corky tissues covering large areas of the tuber surface. This is called russet scab. Pitted scab can also occur where lesions develop up to 1/2 inch deep. These deep lesions are dark brown to black, and the tissues underneath are often straw-coloured and somewhat translucent. More than one of these lesion types may be present on a single tuber.

Although scab symptoms are usually noticed late in the growing season or at harvest, tubers are susceptible to infection as soon as they are formed. Small brown, water-soaked, circular lesions are visible on tubers within a few weeks after infection.

Mature tubers with a well-developed skin are no longer susceptible, but existing lesions will continue to expand as tubers enlarge. Thus disease severity increases throughout the growing season. Scab is most severe when tubers develop under warm, dry soil conditions.

Coarse-textured soils that dry out quickly are therefore more conducive to scab than are fine-textured soils. A few other conditions can be confused with scab. White, enlarged lenticles, which frequently occur on potato tubers harvested from wet soil, can be mistaken for scab.

Usually this condition will disappear when tubers are dried. Patchy russeting, checking, or cracking of tuber surfaces caused by the fungus Rhizoctonia also may be confused with russet scab. A totally different but uncommon disease called powdery scab, caused by the fungus Spongospora subterranea, causes very similar scab-like symptoms. Laboratory examination may be necessary to identify these diseases.

Causal Organisms

Scab is caused by a group of filamentous bacteria called actinomycetes that occur commonly in soil. In soils with a pH above 5.5, Streptomyces scabies is usually responsible for common scab, and is capable of causing all the types of scab lesions. It is commonly introduced into fields on seed potatoes, and will survive indefinitely on decaying plant debris once the soil is contaminated.

Because the organism can survive passage through the digestive tract of animals and be distributed.

BLACKLEG, AERIAL STEM ROT, AND TUBER SOFT ROT OF POTATO

Blackleg, aerial stem rot, and tuber soft rot are all similar diseases caused by several types of soft-rot bacteria. Blackleg and tuber soft rot occur wherever potatoes are grown. Aerial stem rot is also widespread, but is most severe under sprinkler-irrigation.

SYMPTOMS

Blackleg begins from a contaminated seed piece, but the symptoms can occur at several stages of plant development. In severe cases, entire seed pieces and developing sprouts may rot in the ground prior to emergence, resulting in a poor stand. Blackleg often develops after plants are well up or even

in flower. In this case, stem bases of diseased plants typically show an inky-black to light-brown decay that originates from the seed piece and can extend up the stem from less than an inch to more than two feet. Leaves of infected plants tend to roll upward at the margins, become yellow, wilt, and often die. Aerial stem rot (also called bacterial stem rot or aerial blackleg) is initiated by soft-rot bacteria from sources external to the seed piece. Stem infection can occur through wounds or through natural openings such as leaf scars. Lesions on diseased stems first appear as irregular brownish to inky-black areas.

These enlarge into a soft, mushy rot that causes entire stems to wilt and die. Potato tubers with soft rot have tissues that are very soft and watery, and have a slightly granular consistency. The diseased tissue is cream- to tan-coloured, and often has a black border separating diseased from healthy areas. In the early stages, soft-rot decay is generally odorless, but later a foul odor and a stringy or slimy decay usually develop as secondary decay bacteria invade infected tissues. Most internal tuber tissues may be consumed by soft rot organisms, sometimes leaving only a shell of skin remaining in the soil.

Causal Organisms

Blackleg, aerial stem rot, and tuber soft rot are caused by two closely related bacteria, Erwinia carotovora subsp. atroseptica and Erwinia carotovora subsp. carotovora. E. c. carotovora is very common and has an extensive host range, including most fleshy vegetables. It survives readily in soil and surface waters such as rivers, lakes, and even oceans. These bacteria are capable of multiplying and persisting in the root zones of many host and nonhost crop and weed species. In contrast, E. c. atroseptica is associated mostly with potatoes.

These bacteria do not survive well in soil for more than one year, unless they are contained within diseased tubers or other potato plant debris.

Blackleg is usually caused by E. c. atroseptica carried on contaminated seed tubers. Most lots of seed tubers are contaminated to some degree, but the bacteria are usually dormant and do not cause disease unless environmental conditions are favourable.

In contrast, aerial stem rot is usually caused by E. c. carotovora contained in infested soil or introduced to the crop by irrigation water, wind-blown rain, and insects. Tuber soft rot can be caused by either of these soft-rot bacteria. Moisture and temperature are the two critical factors in initiation and development of soft-rot diseases.

High soil temperatures and bruising of seed tubers favour seed-piece decay and pre-emergence blackleg. Blackleg in growing plants is favoured by cool, wet soils at planting followed by high temperatures after emergence. Dense plant canopies and long periods of leaf wetness favour infection of aerial plant parts.

Although tuber soft rot can occur at any temperature above 50 degree F, disease develops best above 75 degree F. Oxygen depletion in tubers also favours soft rot.

When seed pieces in soil or tubers in storage become covered with a film of water, the tissues rapidly become depleted of oxygen. This also may be induced by soil flooding or improper drying of washed tubers. Once it starts, tuber soft rot can proceed rapidly in storage. "Wet" areas may develop in the piled tubers that flow onto ones below, spreading the bacteria. Heat, coupled with condensation on tuber surfaces, can further adversely affect storage conditions, resulting in accelerated "melt" of the pile.

- Plant only certified, disease-free seed tubers. If possible, use whole (B-size) seed tubers that do not have to be cut.

- When receiving seed tubers in bags, do not stack more than five bags high. With bulk or bagged seed, store at 40-45 F until 2-3 weeks before planting, then warm to 55-60 prior to cutting.
- Clean all equipment used for cutting seed tubers thoroughly and then sanitize with an appropriate disinfectant.
- Treat cut seed pieces with recommended fungicide dressings immediately after cutting.
- Plant treated cut seed pieces immediately if soil temperatures are 55-65 degree F at planting depth. Seed pieces can be held 1-2 weeks at 55-60 degree F and 95-99 per cent relative humidity to hasten healing of cut surfaces. Condensation on surfaces of seed pieces must be avoided.
- Do not irrigate fields until plants are well emerged. Avoid using surface water for irrigation.
- During crop growth, monitor irrigation and nitrogen fertility to minimize excessive vine growth that will promote leaf wetness within the plant canopy.
- Harvest tubers only after the vines are completely dead to ensure skin maturity. Low spots in the field should be left unharvested if significant waterlogging has occurred.
- Take all precautions to minimize cuts and bruises when harvesting and handling tubers.
- Hold newly harvested potatoes at 55-60 F with 90-95 per cent relative humidity for the first 1-2 weeks to promote wound healing. After this curing period, lower the temperature of table stock to 38-40 F for long-term storage. Never wash tubers prior to storage.

FUSARIUM DRY ROT AND SEED-PIECE DECAY OF POTATO

Fusarium dry rot is an important postharvest disease of

potato tubers that causes significant losses in storage and transit of both seed tubers and those for table consumption. It is also a major cause of seed-piece decay after planting.

SYMPTOMS

Infected tubers usually develop a dry rot, but a moist rot may occur if secondary infections with soft-rot bacteria also are involved. Surfaces of infected tubers are sunken or wrinkled, and rotted tissues appear brown or gray to black. A white or pink mold is sometimes visible on tuber surfaces. When tubers are cut, internal cavities within rotted tissues may contain white, yellow or pink molds. In storage, blue, black, purple, gray, white, yellow, or pink spore masses may develop in these internal cavities. After low-temperature storage, internal tissues often will become firm and dry or even powdery.

Causal Organisms

Fusarium dry rot is caused by several species of the soilborne fungus Fusarium. These fungi are common in most soils where potatoes are grown and survive as resistant spores free in the soil or within decayed plant tissues. Although some infections may develop on tubers before harvest, most infections occur as the fungus enters tubers through harvest wounds. Small, brown lesions appear at wound sites 3-4 weeks after harvest and continue to enlarge during storage, taking several months to develop fully.

The disease develops fairly rapidly at temperatures above 50 F, but lesions will cease enlarging below 40 F. The fungus is only dormant at these low temperatures, however, and will resume growth when tubers are warmed.

Fusarium seed-piece decay is really the same disease. Seed tubers may be infected prior to shipment. Decay during transit or storage often accounts for poor quality seed tubers. When these Fusarium fungi are present on seed pieces or in the soil, poor stands may result, especially if cut surfaces of seed pieces are not properly healed. Fusarium seed-piece decay begins as reddish-brown to black depressions on cut surfaces.

These may expand to cover the entire seed piece and often result in a slimy rot when infection by secondary soft-rot bacteria follows.

- Harvest tubers only after the vines are completely dead to ensure skin maturity.
- Take all precautions when harvesting and handling tubers to minimize cuts and bruises.
- Hold newly harvested potatoes at 55-60 F with 90-95 per cent relative humidity for the first 1-2 weeks to promote wound healing. After this curing period, lower the temperature of table stock to 38-40 F for long-term storage.
- Plant only certified, disease-free seed tubers. If possible, use whole (B-size) seed tubers that do not have to be cut into seed pieces before planting.
- When receiving seed tubers in bags, do not stack more than five bags high. With bulk or bagged seed, store at 40-45 F until 2-3 weeks before planting. Then allow seed potatoes to warm prior to cutting.
- Treat cut seed pieces with recommended fungicide dressings immediately after cutting.
- Plant treated cut seed pieces immediately or store them at 55-60 F and 95-99 per cent relative humidity to hasten healing of cut surfaces. Condensation on surfaces of seed pieces must be avoided.

RHIZOCTONIA STEM AND STOLON CANKER OF POTATO

Rhizoctonia stem and stolon canker, also called black scurf, is an extremely common problem on potatoes that can result in delayed emergence, reduced stands and poor tuber quality.

SYMPTOMS

Rhizoctonia canker often goes unnoticed until harvest when tubers are found to be covered with small brownish-black fungal bodies (sclerotia) that look somewhat like bits of black soil that will not wash off. These may vary in size from a pinhead to as large as a pea. When introduced into soil, these structures can survive and germinate the following spring to attack young shoots, roots, stolons, and tubers of the new crop.

In many cases, weak, spindly-looking or late-emerging plants may be the result of an attack by this fungus. The first sprouts are often killed before they reach the surface, resulting in the emergence of a weaker secondary sprout. Dry, sunken, brownish lesions developing on the base of the stem below the soil line are evidence of the stem canker phase of this disease.

The lesions may girdle the stem or large cankers may interfere with movement of nutrients from the leaves to the tubers. In such cases, vines become yellow to reddish purple; the leaves begin to curl upward; the stalks swell, particularly at the nodes; and often small purplish tubers form where leaves branch from stems.

During midseason under a dense canopy of foliage, the fungus may develop a white, powdery mold growth on stems, extending just above the soil line. Cankers that form on stolons may prune off young developing tubers. This disease may lead to russeting or surface cracking of mature tubers and sometimes shallow, brown lesions will form around lenticles. Low soil temperatures in the fall favour formation of sclerotia on tubers.

Causal Organism

This disease is caused by the fungus Rhizoctonia solani which can survive in the soil for many years, even under relatively dry conditions. Rhizoctonia causes disease in a wide variety of crops, but the strains found in association with potato generally do not attack and reproduce on other plant species. The fungus survives in soil associated with decomposing plant residues.

In addition, the sclerotia can survive on infected tubers and can persist free in soil for extended periods. Emerging sprouts are usually attacked by fungus present on the seed tubers. Once green leaves develop on sprouts, stem tissues are much less susceptible to infection.

Stem cankers, stolon infections, and sclerotia on tubers usually develop when these tissues grow in proximity to sources of the Rhizoctonia fungus in soil. Cool (55-60 degree F), moist soils are optimal for infection. Sclerotia form on the surfaces of mature tubers under cool, moist conditions, generally after the vines have begun to die.

- Use a crop rotation with corn, grasses, and cereal grains. If this disease has been severe, 3-5 years should elapse between potato crops.
- Plant certified seed tubers that are free of Rhizoctonia on the skin.
- Use planting practices that promote rapid emergence: Avoid planting in heavy, poorly-drained soils. Plant seed potatoes when soil is warm (above 60 degree F). Cover seed tubers with no more than 2 inches of soil.
- Harvest tubers promptly after vines are dead to avoid the development of sclerotia on the surfaces of tubers while still in the soil.

MOSAIC VIRUS DISEASES OF VINE CROPS

Mosaic diseases of vine crops are caused by at least five different viruses:

- Cucumber mosaic virus (CMV),
- Squash mosaic virus (SQMV),
- Watermelon mosaic virus-2 (WMV-2),
- Zucchini yellow mosaic virus (ZYMV)
- Papaya ringspot virus-watermelon isolate (PRSV-W).

All five of these viruses can infect all cultivated vine crops (squash, melons, gourds, cucumbers, pumpkins) and under ideal conditions can cause a high rate of crop failure and severe economic losses.

SYMPTOMS

As the names imply, all five viruses cause a mottling of foliage called mosaic. This is characterized by the presence of intermingled patches of normal and light green or yellowish coloured plant tissue.

Depending on environmental conditions, mosaic symptoms can range from mild to severe and be visible on both leaves and fruit. The younger the plant when infected, the more severe the symptoms as the plant matures. In some cases, plants infected at the seedling stage may collapse and die. Plants infected at the flowering stage may not set fruit or young fruits may abort. If plants are more mature when infected they do not show severe mosaic and may still produce marketable fruit. The most dramatic symptoms are often associated with infected fruit.

Fruit symptoms can range from subtle colour change to severe deformation. It is not uncommon to have two or more viruses infecting the same plant and in these cases symptoms may be much more severe than if the plant were infected with only one virus. It is nearly impossible to distinguish between any of the five viruses based only on visible symptoms.

Because of this, it is important to have suspect plants sent to a diagnostic clinic capable of using specialized techniques to identify the viruses that are present. Proper disease management depends on knowing which specific virus is involved.

Disease Cycle and Transmission

All five vine crop mosaic viruses are transmitted by insects. CMV, ZYMV, WMV-2 and PRSV-W are transmitted primarily by the green peach aphid and the melon aphid. SQMV is transmitted by the striped cucumber beetle and spotted cucumber beetle and also through seed, but only at a very low rate. In most cases these insects acquire the virus by feeding on weeds or other vine crops that are infected.

The insects then feed on noninfected vine crops and in the process of this feeding transmit the virus to the host plant. In the case of the green peach and melon aphid, the insect carries the virus on its probing mouthparts and the aphid only needs to probe the plant to transmit the virus. Feeding by the aphid is not necessary. Once the insect probes one plant, it can move onto the next, probing and infecting as it goes.

The more aphids present, the quicker and more complete

the virus spread. By this process, large acreages can become infected quickly.

- Plant varieties with resistance to these viruses whenever possible. The availability of varieties with virus resistance varies depending on the virus and the vine crop in question. Major seed producers list resistant varieties in their seed catalogs.
- Control weeds which could harbor the viruses from areas around production fields. Pokeweed is a primary weed in the area which harbors many plant viruses. Weeds also serve as hosts for insects that transmit the viruses.
- Since the virus is moved about by insects it would seem logical that insect control would be a primary method of virus control. However, in this case it is not that simple. Since aphid vectors need only probe the plant to transmit the virus, insect control is not really that effective in controlling the virus. Even if the aphid were to die immediately following probing, virus transmission would still take place. To control the insect effectively insecticides would need to be applied on a daily basis. The economics of this may not be feasible. Aphids also move long distances with weather fronts making local control difficult.

DAMPING-OFF AND ROOT ROT OF BEANS

Root rots, damping-off before and after seedling emergence, and seed rots are destructive diseases of green, snap, Lima, and dry beans. These diseases are caused primarily by soil borne fungi.

Significant losses may occur to susceptible varieties, especially if cool, wet weather conditions prevail for the first few weeks after seeding and then are followed by hot, dry weather. Disease incidence and severity often vary greatly, even in areas with a history of root rot.

In the same growing season, it is not uncommon to lose a crop completely and then re-seed and experience no problems. This situation results from changes in biological, environmental, and soil conditions. Since there are no commercially acceptable resistant varieties, growers should learn how to recognize these diseases and use a combination of management practices to minimize potential losses.

SYMPTOMS

Damping-off before emergence results from fungal attack of germinating seeds and/or young seedlings while they are still in the ground. Infected seeds may fail to germinate, become soft and mushy, and finally disintegrate. Slightly darkened water-soaked lesions may be visible on stems of young seedlings. Infected areas enlarge quite rapidly, and seedlings may die shortly after infection, prior to emergence from soil. Roots or stems of seedlings that have already emerged also can be attacked at or below the soil line resulting in damping-off.

Infected roots are usually discoloured or rotted and sometimes reddish-brown lesions develop on the tap root. Infected stem tissues are soft and colourless to dark-brown.

Basal portions of invaded stems may be much thinner than the areas above the lesion, a condition called "wire stem." As a result, the seedling may fall over and die.

Damping-off is a major cause of poor stand establishment in bean plantings. Older plants can also be attacked by these fungi. Later infections are usually confined to roots, which may result in stunting, wilting, or plant death. To diagnose bean root rots, suspected plants should be carefully dug and washed, because pulling plants may leave tissues with characteristic symptoms in the soil. If plants are brought to a diagnostic clinic, they should be dug and left intact in soil.

Causal Organisms

Root rots, damping-off, and seed rots are caused primarily by soilborne fungi. Fusarium root rot is caused by Fusarium solani f. sp. phaseoli. This fungus is capable of surviving long periods in soil, even in the absence of beans, by the production of thick-walled resting spores. Its host is mainly green beans, but lima beans and garden peas are also susceptible. Coarse-textured, acidic and poorly fertilized soils favour development of Fusarium root rot. A number of fungi in the genus Pythium are capable of inciting seed rot and seedling damping-off.

This may vary with environmental conditions and developmental stage of the host. These fungi can survive in soil for many years, either by producing thick-walled resting spores, or in a vegetative condition within crop residues left from previous years. They may also survive by attacking several other garden vegetable crops such as beets, cabbage, peas, melons, squash, and cucumbers. Development of diseases caused by Pythium is more prominent under:

- High or low temperatures that is unfavourable to crop growth,
- Excess water and/or nitrogen,
- Continued cropping of susceptible plants.

Rhizoctonia solani is another soil-inhabiting fungus causing root rots and damping-off. This fungus overwinters free in the soil or within decayed plant tissues. Aside from bean, its host range includes beets, cabbage, lettuce, peas, pepper, tomatoes, and many others. Warm, relatively dry soil conditions favour disease development.

- Do not grow beans or other susceptible crops continually in the same location. Continuous cropping of susceptible plants will eventually lead to a buildup of these fungi in the soil. Since they are capable of long-term survival, a rotation of 4-5 years is desirable. Avoid planting beans in fields known to be heavily infested with bean root-rot fungi.
- Plant beans only on well-drained soils or try to improve drainage. This could be done by improving soil structure and/or installing drain tiles. Subsoiling to a depth below the plowed layer will reduce soil compaction, and improve drainage. Deep plowing of the previous years' crop residues will reduce bean root rot.
- Delay planting until the soil is warm (above 65 F) and seed shallow to insure rapid emergence. Avoid planting seeds too close together-follow instructions

on the seed container. Do not overfertilize, especially with nitrogen.

- Use of fungicide-treated seeds will minimize problems with damping off and root rots.

POWDERY MILDEW OF VINE CROPS

Powdery mildew can be a serious problem on foliage and stems of cucumbers, melons, squash, pumpkins, and ornamental gourds.

SYMPTOMS

Small patches of fine, white threads develop on surfaces of infected leaf blades in late July on most vine crops. These patches grow together and eventually cover stems and foliage with white, powdery masses of spores.

In severe infections, leaves will yellow and die. Petioles, stems and, rarely, fruit will also become infected. Early death of leaves can decrease the total amount of fruits formed as well as reduce fruit size. Loss of foliage hastens maturity of fruit and increases sunburning. Stems infected with Powdery Mildew become brown. This can spoil the appearance of the "handles" on pumpkins which lowers value of the fruit.

Causal Organisms

Two fungi cause powdery mildew on vine crops, Sphaerotheca fuliginea and Erysiphe cichoracearum. S. fuliginea is most common. Powdery mildew fungi thrive under conditions of high relative humidity, warm temperatures, low light, high fertility, and succulent plant growth. Unlike bacteria and many fungi, free moisture on leaf surfaces actually inhibits infection by these pathogens, although very high relative humidity is required for spore germination. Powdery mildew fungi grow only on living host plants.

They survive the winter as dormant mycelium on perennial plants or as spores in thick-walled fruiting structures. Powdery Mildew spores can be blown into the spring from warmer southern areas. Mature foliage is most readily infected; very young leaves are nearly immune.

Control

- Plant resistant cultivars of cucumbers and muskmelons. Pumpkin varieties vary greatly in susceptibility to this disease.
- Use cultural practices that avoid excessive succulence, overcrowding, shading, overwatering, or excess fertilization especially with nitrogen.
- Avoid making new planting of vine crops in the vicinity of older plantings, especially if mildew is already present.
- Beginning in late July or early August when Powdery Mildew first appears, regular applications of fungicides may be required.

PHYTOPHTHORA BLIGHT OF PEPPER AND CUCURBITS

Phytophthora blight, a highly destructive disease of peppers and cucurbits. It can become a serious problem during periods of heavy rainfall; the pathogen can spread rapidly through the crop, resulting in severe losses within a short time.

SYMPTOMS

Peppers

Phytophthora blight affects both seedlings and mature plants. Infected seedlings show typical damping-off symptoms. Infection of older plants usually begins at or below the soil line.

Water-soaked, dark brown lesions on the lower stems (collar rot phase) usually extend upward for an inch or more above the soil line and may expand to girdle the stems, preventing upward movement of water and nutrients. This often results in a sudden wilting of foliage. Root infections may also occur which kill roots and cause wilting of the plant without the appearance of stem cankers.

The foliar phase of this disease commonly occurs at forks in the branches, resulting in dark, girdling cankers and wilting of leaves and fruits. Infected leaves develop circular or irregular, dark green, water-soaked lesions which dry and appear light tan. Fruit lesions may also appear as enlarging, watersoaked areas, which then shrivel and darken. A mass of white fungal growth may develop inside the fruit, and seeds usually turn dark brown or black. A fine, grayish-white to tan mold may also become evident over the lesion on the fruit surface. Under humid conditions, fungal growth develops extensively over the entire fruit.

Cucurbits

Seedlings show typical symptoms of damping-off. Mature plants may wilt suddenly as a result of root infection, in the absence of obvious stem or vine lesions. Stem and leaf petiole lesions are light to dark brown, watersoaked and irregular, eventually becoming dried and papery. Fruit symptoms begin as small, water-soaked circular lesions, which expand to become large, soft, sunken areas covered with white fungal growth. Infected fruit often collapse or "melt down" in the field or in storage.

CAUSAL ORGANISMS

Phytophthora blight is caused principally by the soil-borne fungus, Phytophthora capsici, although a similar fungus, Phytophthora parasitica, has also been reported to cause fruit rot on peppers and cucurbits. Both of these fungi are dependent on free water in soil for infection. Because of this, initial infection by Phytophthora usually occurs on plants growing in poorly drained areas of fields.

The fungi that cause Phytophthora blight survive as thick-walled, resistant spores (oospores) in the soil and as mycelium in infected plant tissues. Once introduced, these fungi can survive up to 15 months in moist soil in the absence of host plants. They can also be carried on seed or transplants.

Once stem infection occurs, the fungi produce spores on infected stem tissues, which are then carried by splashed rain onto nearby plants.

Lower branches of adjacent plants can also be infected by rain-splashed soil contaminated by run-off water. Fruits in contact with the soil are especially prone to infection. Spores are produced on newly infected fruit and stems, and new infections can develop quickly in a short time. Warm (68 to 75 degrees F), wet weather is most favourable for disease development and spread of the pathogen. Disease progress declines when dry weather returns.

- Use only certified disease-free seed or transplants.
- Produce pepper plants on raised beds to retard initial stem infection. Arrange beds to prevent puddle formation between rows. Avoid depressions on the surface of the beds where water can accumulate, especially around the base of plants. Avoid planting peppers or cucurbits in poorly drained fields.
- Practice crop rotation so that peppers or cucurbits are grown only every 3 to 4 years to reduce the amount of Phytophthora present in the soil. Use non-susceptible crops such as corn, small grains, beans, crucifers, or potatoes in the rotation.
- Application of fungicides may reduce disease development and spread if used in combination with cultural practices such as crop rotation and raised beds. Application at planting and two or more (depending on the fungicide used) applications after planting are required.

BLOSSOM-END ROT OF TOMATO, PEPPER, AND EGGPLANT

Blossom-end rot is a serious disorder of tomato, pepper, and eggplant. Growers often are distressed to notice that a dry sunken decay has developed on the blossom end of many fruit, especially the first fruit of the season. This nonparasitic disorder can be very damaging, with losses of 50 per cent or more in some years.

SYMPTOMS

On tomato and eggplant, blossom-end rot usually begins as a small water-soaked area at the blossom end of the fruit. This may appear while the fruit is green or during ripening. As the lesion develops, it enlarges, becomes sunken and turns black and leathery. In severe cases, it may completely cover the lower half of the fruit, becoming flat or concave.

Secondary pathogens commonly invade the lesion, often resulting in complete destruction of the infected fruit. On peppers, the affected area appears tan, and is sometimes mistaken for sunscald, which is white. Secondary molds often colonize the affected area, resulting in a dark brown or black appearance. Blossom end rot also occurs on the sides of the pepper fruit near the blossom end.

Cause

Blossom-end rot is not caused by a parasitic organism but is a physiologic disorder associated with a low concentration of calcium in the fruit. Calcium is required in relatively large concentrations for normal cell growth. When a rapidly growing fruit is deprived of necessary calcium, the tissues break down, leaving the characteristic dry, sunken lesion at the blossom end. Blossom-end rot is induced when demand for calcium exceeds supply.

This may result from low calcium levels or high amounts of competitive cations in the soil, drought stress, or excessive soil moisture fluctuations which reduce uptake and movement of calcium into the plant, or rapid, vegetative growth due to excessive nitrogen fertilization.

- Maintain the soil pH around 6.5. Liming will supply calcium and will increase the ratio of calcium ions to other competitive ions in the soil.
- Use nitrate nitrogen as the fertilizer nitrogen source. Ammoniacal nitrogen may increase blossom-end rot as excess ammonium ions reduce calcium uptake. Avoid over-fertilization as side dressings during early fruiting, especially with ammoniacal forms of nitrogen.
- Avoid drought stress and wide fluctuations in soil moisture by using mulches and/or irrigation. Plants generally need about one inch of moisture per week from rain or irrigation for proper growth and development.

- Foliar applications of calcium, which are often advocated, are of little value because of poor absorption and movement to fruit where it is needed.

CLUBROOT OF CRUCIFERS

Clubroot is a world-wide problem in temperate climates in the production of cruciferous vegetables such as cabbage, broccoli, cauliflower, radishes, and turnips; and field crops such as mustard and rape. The disease was known as early as the 13th century in England where it was called "finger and toe" disease because of thc shape of infected roots.

SYMPTOMS

The most striking symptom of clubroot is an abnormal enlargement of the root system with clubs often thickest at the centre, tapering spindle-like towards the ends. In radishes, clubroot causes distorted swellings on the base of the bulb and along the tap root. In severe cases, entire plantings are destroyed. Clubroot-infected plants often wilt on sunny days and permanent wilting may accompany advanced decay of infected roots. Severe stunting may be evident if infection occurs early and the disease progresses rapidly. The malformed and greatly enlarged roots are the key symptom of this disease.

Causal Organism

Clubroot is caused by the soil-borne fungus, Plasmodiophora brassicae, which only infects plants in the crucifer family. It infects susceptible host plants through root hairs. Once in the tissue, it stimulates abnormal growth of affected parts, resulting in a swollen club. Infection is favoured by excess soil moisture and low pH, although it can occur over a wide range of conditions. Once a plant is infected, numerous resistant spores of the fungus are produced in the "clubbed" tissues.

As these tissues decay, spores are released into the soil where they can remain infectious for at least 10 years. Contaminated soil moved by wind or water can serve as a source of infestation of nearby fields causing outbreaks of disease in areas where susceptible crops are planted for the first time. Numerous races of the pathogen have been identified.

Clubroot is a very difficult disease to manage, and heavily infested areas may have to be abandoned for crucifer production. Some control may be achieved with the following measures:

- A good crop rotation programme, growing crucifers on the same soil no more than every third or fourth year, is essential to retard development of a large population of spores on land not already heavily infested.
- Since clubroot is favoured by a low pH, liming soil to pH 7.2 or above may be helpful. Raising soil pH too high, however, may interfere with the growth of succeeding crops other than crucifers. Calcitic lime is usually preferable to dolomitic lime, except for soils low in magnesium, where dolomitic lime is more effective. In course-textured soils, increasing the pH can result in boron deficiency. This may be alleviated by application of boron in transplant water or as a foliar spray.
- With transplanted crops, the use of pathogen-free seedbeds and uninfected plants is essential to prevent introduction of the disease.
- Application of a fungicide in transplant water or rototilled in a band prior to planting may help to reduce disease development.
- Clean and disinfect all machinery before moving it from infested to non-infested land.
- Some resistant cultivars are available. However, plant resistance has not been very useful in clubroot control because of rapid development of new races of the fungus.

COMMON SMUT OF CORN

Corn smut is an extremely common disease of sweet, pop,

and dent corn and throughout the world. It is usually not economically important, although in some years yield losses in sweet corn may be as high as 20 per cent. In Mexico, immature smut galls are consumed as an edible delicacy known as cuitlacoche, and sweet corn smut galls have become a high value crop for some growers in the NE United States who sell them to Mexican restaurants.

SYMPTOMS

The corn plant may be infected at any time in the early stages of growth, but becomes less susceptible after formation of the ear. Above-ground parts may be infected, but it is more common to see smut galls on the ears, tassels, and nodes than on the leaves, internodes, and aerial roots. The smut gall is composed of a great mass of black, greasy, or powdery spores enclosed by a smooth white covering of corn tissue.

The gall may be 4-5 inches in diameter. When leaves are infected, small pustules develop, usually on the midrib, causing some leaf distortion. After the spores mature, the outer covering becomes dry and brittle, breaks open, and the spores sift out. Greatest yield losses occur when the ear becomes infected or if smut galls form on the stalks immediately above the ears.

Causal Organism

Corn smut is caused by the fungus, Ustilago zeae, that survives as a resistant spore in the soil over winter, and possibly for 2 to 3 years. These spores, called teliospores, can be blown long distances with soil particles or carried into a new area on unshelled seed corn and in manure from animals that are fed infected corn stalks. The teliospores germinate in moist air and give rise to tiny spores called sporidia. The sporidia bud like yeast, forming new spores that germinate in rain water that collects in the leaf sheaths.

This leads to infections that are visible in 10 days or more. Wounds from various injuries provide points for the fungus to enter the plant.

The smut fungus is sensitive to temperature and moisture

changes. In a warm season, the amount of smut is related closely to the amount of soil moisture, especially during June. When temperatures are lower than normal, there may be little smut even though soil moisture remains high.

- Seed treatment is of no value because few spores are on the corn seed.
- For the home garden, remove smut galls before they break open and bury or burn them. This must be done on a community basis, however, in order to be effective. Smut gall removal is not practical in commercial production.
- In order to reduce infection points from insect injury, control corn borers as first tassels appear by application of insecticides when insect populations are high.
- Avoid injury of roots, stalks, and leaves during cultivation.
- Deep plow diseased corn stalks in the fall to bury surviving spores.
- Use resistant hybrids or varieties. Dent corn is generally more resistant than sweet or popcorn. In sweet corn, the larger, later-growing varieties usually are more resistant than the smaller, early varieties.

BACTERIAL SPOT, SPECK, AND CANKER OF TOMATOES

Bacterial spot, bacterial speck and bacterial canker are widespread diseases of tomato that can cause localized epidemics during warm (spot and canker) or cool (speck), moist conditions. Bacterial spot can cause moderate to severe defoliation, blossom blight, and lesions on developing fruit. Bacterial speck also causes these symptoms but is usually not as severe as bacterial spot. Bacterial canker causes wilt, vascular discolouration, scorching of leaf margins, and lesions on fruit.

SYMPTOMS

Foliar symptoms of bacterial spot and speck are identical.

Small, water-soaked, greasy spots about 1/8 inch in diameter appears on infected leaflets. After a few days, these lesions are often surrounded by yellow halos and the centres dry out and frequently tear.

Lesions may coalesce to form large, irregular dead spots. In mature plants, leaflet infection is most concentrated on fully-expanded and older leaves and some defoliation may occur. Spots may also appear on seedling stems and fruit pedicels. In some cases, blossom blight may occur, causing flower abortion.

This is more severe with bacterial spot and may result in a split fruit set which is especially troublesome with determinate cultivars intended for mechanical harvest. Bacterial spot and speck can usually be differentiated by symptoms on immature fruits. Bacterial spot lesions are small water-soaked spots that become slightly raised and enlarged until they are about 1/4 inch in diameter. The centres of these spots later become irregular, light brown, slightly sunken with a rough, scabby surface. In the early stages of infection, a white halo may surround each lesion at which time it resembles the fruit spot of bacterial canker.

Small lesions which have not yet become scabby are often confused with lesions of bacterial speck. Bacterial speck appears on immature fruit as a black, slightly sunken stippling, eventually causing lesions less than 1/16 inches in diameter. Fruit lesions are not initiated on mature fruit in either disease. Primary or systemic symptoms of bacterial canker (from infections originating in seeds or young seedlings) include stunting, wilting, vascular discolouration, development of open stem cankers, and fruit lesions.

When affected stems are split open lengthwise, a thin, reddish-brown discolouration of the vascular tissue is observed, especially at the base of the plant. On young seedlings in the greenhouse, lesions may appear as raised pustules on leaves and stems. These plants rarely survive the season in the field.

Secondary symptoms in the field include leaf "firing" (necrotic marginal leaf tissue adjacent to a thin band of chlorotic tissue) and fruit lesions. Spots on fruit are relatively small (1/32 to 1/16 inch) surrounded by a white halo ("bird's-eye" spots). Canker bacteria may also invade internal fruit tissues, causing a yellow to brown breakdown.

Causal Organisms

Bacterial spot is caused by the bacterium, Xanthomonas campestris pv. vesicatoria, which can be carried as a contaminant on the surface of infested seed and has been found to overwinter in soil associated with plant debris. Bacterial speck is caused by another bacterium, Pseudomonas syringae pv. Tomato.

This bacterium may also be seedborne and can overwinter on plant debris in soil and on the roots of many perennial plants. Bacterial canker is caused by Clavibacter michiganensis subsp. michiganensis, which, unlike the spot and speck pathogens, has the ability to infect tomato plants systemically.

It is seedborne and can survive on infested plant debris in soil. All three organisms may exist at low populations on leaf surfaces of symptomless plants. At the onset of favourable conditions, these low populations can increase rapidly and bacteria can then enter plants through stomata or small wounds and begin infection.

Bacteria can spread rapidly with spattering rain and widespread epidemics may develop. Penetration of tomato fruit occurs through wounds created by windblown sand, breaking of hairs, or by insect punctures.

Optimal conditions for bacterial spot and canker are high moisture, high relative humidity and warm temperatures (75 to 90 degrees F). Bacterial speck is more likely to occur under cool (64 to 75 degrees F), moist conditions.

- Rotate tomatoes with non-solanaceous crops with at least 2 to 3 years between tomato crops. Avoid rotation with peppers, which are also susceptible to bacterial spot.
- Plant only seed from disease-free plants or seed treated to reduce any bacterial populations. Treatments include:

a. Fermentation of tomato pulp and seeds at room temperature for 4–5 days;
b. Soaking seeds in 0.6–0.8 per cent acetic acid for 24 hr at 70 degrees F;
c. Soaking seeds 5ñ10 hr in 5 per cent hydrochloric acid;
d. Hot water treatment of seeds (122 degrees F for 25 minutes); or
e. Sodium hypochlorite (bleach) treatment [20–40 minute soak of seeds in 1 per cent sodium hypochlorite (20 per cent bleach)]. Some decrease in germination may be expected from these treatments.

- Use only transplants free of disease symptoms.
- Carry out proper sanitation of transplant production greenhouses. Remove all weeds and plant debris, clean all tools with disinfectant solution, and wash hands thoroughly before and after handling plants. Water plants early in the day to reduce the amount of time foliage is wet.

  Do not handle plants when they are wet. After each crop, clean greenhouse walls, benches, etc., with hot soapy water, followed by thorough rinsing and treatment with a disinfectant. If possible, close up greenhouse after transplant production is completed to allow natural heating during the summer. Use only new plug trays and pathogen-free planting mixes. Avoid growing peppers and tomatoes in the same greenhouse unless pepper seed has also been treated as in step
- In the field, control irrigation to minimize the time foliage is wet and avoid working among wet plants to minimize spread of disease.
- Applications of mancozeb plus copper soon after transplanting may help retard development and spread of bacterial spot and speck. This practice is not particularly effective for management of bacterial canker. Many tomato processors will not accept

tomatoes treated with mancozeb or other EBDC fungicides.

BACTERIAL WILT OF CUCURBITS

After vine crops begin to run, gardeners and farmers often notice individual leaves with severe wilt symptoms on sunny days. Within a week or two the condition spreads to entire vines which do not recover from the wilt. This disease, called bacterial wilt, is especially common with cantaloupes and cucumbers. Squash and pumpkins may not wilt as rapidly, but may be dwarfed with extensive blossoming and branching. Watermelons are rarely affected.

SYMPTOMS

Wilting of individual leaves or vines of the plant is the characteristic symptom. One or a few leaves wilt and become dull green. The disease spreads from the leaves downward into the petioles and then the stem until the entire plant wilts and dies. There are other factors, such as vine borers and soil-borne fungal pathogens that may cause cucurbits to wilt.

Sometimes, if an affected stem is cut off near the ground, the sap may be milky in appearance or sticky and, if touched with the finger, the sap will string up to half an inch. This is a helpful test in diagnosis of bacterial wilt, but cannot be depended upon for positive identification.

Causal Organism

This disease is caused by a bacterium, Erwinia tracheiphila that overwinters in the bodies of the striped and 12-spotted cucumber beetles. In the spring, the beetles emerge from the ground and feed on young plants, introducing bacteria into the leaves or stems.

The bacteria reproduce in the water-conducting vessels, producing gums that interfere with water transport. The beetles and bacteria are so intimately related that controlling the beetles will control infection by the bacteria. Once infection has occurred, however, no control is possible and wilting plants should be removed, if practical.

Management

The only practical management measure is to use an insecticide when seedlings first emerge to control the black and yellow cucumber beetles. Early infections are most severe, but total control depends on applications continuing at frequent intervals during the growing season.

In some cases, if insect pressure is heavy, it may be necessary to apply an insecticide when plants are just cracking the soil, but have not yet emerged. Management of this disease is completely linked with preventing feeding of cucumber beetles on susceptible hosts.

FUSARIUM AND VERTICILLIUM WILTS OF TOMATO, POTATO, PEPPER, AND EGGPLANT

Solanaceous crop plants (tomato, potato, pepper, and egg-plant) may be infected at any age by the fungi that cause Fusarium wilt and Verticillium wilt. The wilt organisms usually enter the plant through young roots and then grow into and up the water conducting vessels of the roots and stem. As the vessels are plugged and collapse, the water supply to the leaves is blocked. With a limited water supply, leaves begin to wilt on sunny days and recover at night.

Wilting may first appear in the top of the plant or in the lower leaves. The process may continue until the entire plant is wilted, stunted, or dead. Tomato and potato plants may recover somewhat but are usually weak, unthrifty, and produce fruit of low quality.

Peppers typically collapse rapidly and die. Fusarium and Verticillium wilts are rarely significant in field grown tomatoes due to the widespread incorporation in tomato cultivars of genes for resistance to the pathogen. However, the resurgent interest in planting "heirloom" tomato varieties which do not carry resistance genes has resulted in increased incidence of Fusarium and Verticillium wilts.

Additionally, new races of both pathogens have been identified that are capable of overcoming the resistance in many popular tomato varieties. Verticillium race 2 is now common in tomatoes, but its importance in reducing yield is not known at this time. There is very little genetic resistance available to either disease in pepper or eggplant.

SYMPTOMS

Fusarium Wilt

Fusarium wilt symptoms begin in tomato and potato as slight vein clearing on outer leaflets and drooping of leaf petioles. Later the lower leaves wilt, turn yellow and die and the entire plant may be killed, often before the plant reaches maturity.

In many cases a single shoot wilts before the rest of the plant shows symptoms or one side of the plant is affected first. If the main stem is cut, dark, chocolate-brown streaks may be seen running lengthwise through the stem.

This discolouration often extends upward for some distance and is especially evident at the point where the petiole joins the stem. Potato tubers may show browning of the vascular ring as well as browning at the stem end and decay where stolons are attached. In pepper, lower leaves do not begin to wilt until roots and the base of the stem have already started to decay. Wilting of the entire plant soon follows. Dark brown, sunken, and eventually girdling cankers may be seen at the base of the pepper plant. In eggplant, wilting progresses from lower to upper leaves, followed by collapse of the plant.

Verticillium Wilt

Verticillium wilt symptoms on tomato, potato, and eggplant are similar to those of Fusarium wilt. Often no symptoms are seen until the plant is bearing heavily or a dry period occurs.

The bottom leaves become pale, then tips and edges die and leaves finally die and drop off. V-shaped lesions at leaf tips are typical of Verticillium wilt of tomato. Infected plants usually survive the season but are somewhat stunted and both yields and fruits may be small depending on severity of attack. A light tan discolouration in the stem similar to that caused by Fusarium wilt may be found but is usually confined to lower plant parts. The discolouration is typically lighter in colour than with Fusarium wilt. Symptoms on one side of the plant only are sometimes seen. In potatoes the pathogen may be part of a complex that includes, among others, the root lesion nematode and the bacterial soft rot organism, resulting in premature plant death ("potato early dying disease").

Tubers from Verticillium-infected plants may show light brown vascular discolouration, usually restricted to the stem end. In pepper, the lower leaves wilt, then leaf tips and margins dry and turn brown. Brown streaks in the vascular tissue can be observed well up into the plant, which rapidly collapses and dies.

Causal Organisms

Fusarium wilt in solanaceous crops is caused by several different types of the fungus Fusarium oxysporum. These are: F. oxysporum f. sp. lycopersici (tomato), F. oxysporum f. sp. melongenae (eggplant) and F. oxysporum var. vasinfectum (pepper). Fusarium wilt in potato is caused by a complex of up to four different Fusarium spp. All of the Fusarium wilt pathogens are generally specific to their hosts and are soilborne. They are warm weather organisms, and therefore Fusarium wilts are most serious later in the growing season.

Verticillium wilt is caused by the fungi Verticillium albo-atrum and V. dahliae. These fungi attack a wide range of plant species, including cultivated crops and weeds.

They are soilborne in field and greenhouse soils where they can persist for many years. V. albo-atrum is a cool weather organism that grows best when soil temperatures are between 65 and 75 degrees F. V. dahliae is more active between 75 and 83 degrees F. Although disease is retarded by the higher temperatures that favour Fusarium wilt, visible symptoms may appear to be more severe when high temperatures exist, due to restricted water movement in the plant brought about by damage done to the water conducting vessels earlier in the growing season.

- Because Fusarium and Verticillium fungi are widespread and persist several years in soil, a long crop rotation (4 to 6 years) is necessary to reduce populations of these fungi. Avoid using any solanaceous crop (potato, tomato, pepper, eggplant) in the rotation, and if Verticillium wilt is a problem, also avoid the use of strawberries and raspberries, which are highly susceptible. Rotate with cereals and grasses wherever possible.

- Keep rotational crops weed-free (there are many weeds hosts of Verticillium).
- Whenever practical, remove and destroy infested plant material after harvest.
- Maintain a high level of plant vigour with appropriate fertilization and irrigation, but do not over-irrigate, especially early in the season.
- Plant disease resistant tomato varieties, labeled V (for Verticillium) and F (for Fusarium). These disease resistance designations are usually shown in seed catalogues. Fusarium- or Verticillium-resistant varieties of eggplant, potato, and pepper are generally not available.
- If soils are severely infested, production of solanaceous crops may not be possible unless soil fumigation is an option.

BACTERIAL SPOT OF PEPPER

Bacterial spot of pepper is one of the most destructive diseases of pepper in climates where high temperature and frequent rainfall occur during the growing season. The disease causes spots on leaves and fruit; leaf defoliation; and a reduction in plant growth, fruit yield, and quality. Bacterial spot is also a serious problem on tomato, although not all strains of the pathogen can cause disease on both hosts.

SYMPTOMS

Characteristic bacterial spot symptoms can appear on the leaves, fruits, stem, and petioles. On leaves, symptoms begin as small, yellow-green circular lesions surrounded by a yellowish halo. These spots appear water-soaked under wet conditions. As the lesions mature, a general yellowing extending from the area around the lesions develops on diseased leaves and the centre of the spots become brown to black and sunken. Tissue in the centre of the lesion often dries and breaks away, giving a "shot-hole" appearance to the leaf.

When spots are numerous, they may join together and form irregular discoloured streaks along the veins and leaf margins. Edges and tips of leaves may die, then dry and break away, causing leaves to appear ragged. Severely spotted leaves turn yellow or brown and fall from the plant; young leaves can be distorted. Fruit spots begin as green, circular, slightly raised lesions which eventually become brown or dark, raised, and about 1/8 inch in diameter.

Centres of the spots become necrotic, corky, and scab-like. On stems and petioles, lesions are elongated and blackened, and can kill leaflets. Cotyledons are particularly susceptible to bacterial spot; lesions are initially small, sunken, and silvery. They later become darker in colour.

Causal Organism

Bacterial spot of pepper is caused by a bacterium, Xanthomonas campestris pv. vesicatoria, which can overwinter in crop residue in or on soil, on or in seeds, and on wild host plants.

Pepper seeds infested with the pathogen are a major source of inoculum for bacterial spot as well as the major means of long distance spread of the pathogen. The pathogen can survive on dried seeds for up to 10 years. It is not able to survive free in soil for a long time, but can survive up to 6 months in infected crop debris in soil.

The bacterium penetrates leaves through stomata and/ or wounds, and fruits through wounds created by wind-driven sand, insect punctures, or mechanical injury. Dissemination of the bacterium occurs between and within fields by water-splashing, aerosols, or during cultivating, hoeing, thinning of direct seeded plants, transplanting, or harvesting.

In the United States, numerous physiological races of the bacterium are known, most of which have been found.

- Use pathogen-free seeds and transplants.
- Use sodium hypochlorite-treated seed to reduce bacterial populations.
- Practice crop rotation with non-host plants such as corn and soybean so that peppers are grown only every 3 to 4 years. However, do not use soybeans in the rotation if white mold (Sclerotinia sclerotiorum) has been a problem.
- Deep plow to bury infected crop debris.
- Avoid working in the field when foliage is wet.
- Eliminate wild host plants such as nightshade and ground cherry in and around field.
- Application of copper-containing pesticides may be helpful for preventing development and spread of bacterial spot.
- None of the currently available pepper varieties are resistant to all known races of the bacterial spot pathogen. However, use of varieties resistant to one or more races of the pathogen may provide some control, depending on the races present.

DOWNY MILDEW OF CRUCIFERS

Downy mildew affects all cultivated plants and weeds in the crucifer family. It can be a serious problem in commercial production of cabbage, broccoli, cauliflower, radish, turnip, mustard, collard, and cruciferous greens. Under favourable conditions, it may cause serious losses in the field or may develop after harvest and cause deterioration of product quality during packing and shipping.

SYMPTOMS

Plants can be infected at any stage of development. In seed beds, cotyledons and primary leaves are invaded resulting in fungal growth visible on the underside of the leaf. Later a slight yellowing develops opposite the fungal growth on the upper side of the leaf.

The young leaf or cotyledon, when yellow, may drop off. Older leaves usually persist and infected areas gradually enlarge, turn bright yellow, then become tan and papery. Rarely the affected leaf may develop hundreds of minute darkened specks.

Under cool, moist conditions, a white mildew growth can be seen on the underside of infected leaf lesions. Symptoms may appear on other plant parts as well. The fleshy roots of turnips and radishes may develop an internal, irregularly shaped discolouration extending from the crown downward. The flesh may be brown or black or show a form of net necrosis. In advanced stages, the skin becomes roughened by minute cracks and the root may split open. In radish these symptoms may be confused with those caused by Rhizoctonia.

Causal Organism

The fungus causing downy mildew in crucifers, Peronospora parasitica, overwinters in roots or in decaying portions of diseased plants. Thick-walled resting spores may form in stems, cotyledons, and other fleshy parts of infected host plants. On growing plants, the fungus produces large numbers of spores that are blown about by wind and splashed by rain. Moisture and temperature are important in the spread and reproduction of this fungus.

High relative humidity during cool or warm, but not hot, periods promotes its growth and sporulation. Presence of a water film on the foliage from fog, drizzling rain, or dew allows spores to germinate, infect, and produce more spores on a susceptible host in as few as 4 days.

- Use a crop rotation plan that excludes production of any type of cruciferous crop for at least 2 out of every 3 years.
- Practice sanitary measures such as the use of clean seed beds away from other crucifer production and the destruction of cruciferous weeds.
- Use a planting site and plant spacing pattern that expose plants to full sun throughout the day.
- If severe disease pressure is expected, apply a registered fungicide weekly beginning soon after emergence.

- Disease resistant cultivars are not available for most cruciferous crops. However, some hybrid cultivars of broccoli are resistant or tolerant to downy mildew.

BLACK ROT OF CRUCIFERS

Black rot is the most serious disease of crucifer crops world wide when environmental conditions (relatively high temperature and humidity) are favourable. The disease affects primarily aboveground parts of plants at any stage of growth and causes high yield and quality losses, especially in tropical and subtropical regions during the rainy season. All vegetables in the crucifer family, including broccoli, Brussels sprouts, cabbage, cauliflower, Chinese cabbage, kale, mustard, radish, rutabaga, and turnip, are susceptible to black rot.

Many cruciferous weeds such as Shepherd's Purse, wild mustard, and yellow rocket may also be hosts of this pathogen. The characteristic black rot symptom on most cultivated crucifer plants is the appearance of yellow, V-shaped lesions along the margins of leaves.

The point of the V-shaped lesion is directed toward a vein. When lesions enlarge, wilted tissue expands toward the base of leaves. Eventually the diseased areas become necrotic and the veins turn black or brown.

The infection may move down the vascular tissue of petioles and then spread up and down the stems. When stems and petioles of an infected plant are cut crosswise or lengthwise, the black-brown vascular tissue with yellowish bacterial slime is observed.

These symptoms may be confused with Fusarium yellows, except that Fusarium causes brown vein discolouration without bacterial slime.

Moreover, symptoms of black rot may vary according to age of host, host genus, species, and cultivar and even environmental conditions. Symptoms on cauliflower may appear as numerous black or brown specks, scratched leaf margins, black veins, and discoloured curds. Many cruciferous weed species do not exhibit any of these characteristic symptoms even when infected.

SYMPTOMS

Causal Organism

Black rot of crucifer is caused by a bacterium, Xanthomonas campestris pv. campestris (Xcc). The bacteria can overwinter in plant debris, in and on seeds from diseased plants, and in and on weeds.

The pathogen may survive in diseased crop residue buried in soil for up to 2 years, but not more than 60 days free in soil. The major source of these bacteria is infected seeds, which enable long-distance spread of the disease. The pathogen is spread within and between fields by splashing water, wind, insects, machinery, and irrigation or drainage waters.

The bacteria infect the cotyledons and young leaves through natural plant openings (stomata, hydathodes) or wounds and then migrate between cells until they reach the xylem tissue where they spread throughout the plant. Free moisture is required for infection by the pathogen. After infection, symptoms may appear on plants within 7 to 14 days under optimum conditions (25 to 30 degrees C).

Effective management of black rot of crucifers depends on the application of the following practices in combination:

- Use black rot-tested, disease-free seed grown in an arid production area.
    a. If source of the seeds is unknown, or infested seedlots must be used, treat seed with hot water to eradicate pathogenic bacteria. Cabbage, broccoli, and Brussels sprouts can be treated at 50 degrees C for 25 minutes, while seeds of cauliflower, kale, turnip, and rutabaga are treated for 15 minutes. However, this treatment may reduce the viability of seed. Therefore, some other chemical seed treatments, including, sodium hypochlorite, hydrogen peroxide, and hot acidified cupric acetate or zinc sulfate can be applied to eliminate the bacteria from crucifer plant seeds.
- Use certified disease-free transplants.

- Practice crop rotation where crucifers are grown only every 3 to 4 years to eliminate the inoculum sources from diseased crop debris in the soil.
- Good sanitation practices should be performed to prevent disease spread.
  a. Eliminate all volunteer crucifer plants from previous crops and alternative wild host plants within and around the field.
  b. Do not apply manure that may contain crucifer residues.
  c. Do not use sprinkler irrigation.
  d. Avoid working in the field when plants are wet.
  e. Do not allow machinery and equipment movement from infested areas to non-infested fields.
  f. Deep plow to bury all crucifer residues after harvest.
- Application of fixed copper pesticides in the field may help to reduce spread of the disease.
- A few black rot-resistant cultivars of cabbage and other crucifers are commercially available. These resistant cultivars should be used in crucifer growing regions where black rot is a common problem.

BLACK ROT OF CUCURBITS

Gummy stem blight is an important disease of squash, pumpkins, cucumbers, watermelons and other field-grown cucurbit crops. It can occur at any growth stage, from seedlings to mature plants. This disease on fruits, in the field or in storage, is called black rot. The disease also can cause extensive damage to all above-ground parts of greenhouse-grown cucumbers.

SYMPTOMS

Gummy stem blight occurs on all plant parts except roots. Leaf symptoms appear as dark yellow or reddish-brown lesions in various shapes. Lesions begin at leaf margins and extend rapidly back into the leaf blade, causing curling, shriveling, and death of the entire leaf. Pimple-like structures (pycnidia) may be found in leaf lesions by close inspection with a hand lens. Fruit symptoms vary among crops. Winter squash (Hubbard, butternut, etc.) are likely to show symptoms primarily on fruit or older leaves.

Black rot symptoms appear as a brown to black rot of the rind, flesh, and seed cavity accompanied by heavy white and black fungus growth. Lesions may develop anywhere on the fruit, first as water-soaked areas dotted with pycnidia that ooze yellowish masses of spores.

On Hubbard-type squash, the brownish-black rot extends down into the flesh and seed cavity. Seeds often become dotted with small black pycnidia. On butternut squash, lesions are brown with irregular ring patterns and are superficial over the skin surface, not penetrating into the flesh or seed cavity. Infection usually occurs in the field, causing water-soaked, cracked, brownish cankers on the vines.

A reddish gum may develop in these cracks, although this alone is not a diagnostic sign. (Fusarium and scab may also produce a reddish gum.) Fruit may decay at the site of attachment as a result of the fungus invading the stem. Butternut squash and gourds may develop black rot before harvest, but Hubbard squash are resistant during growth and do not show symptoms until the storage period. Fruit rot on greenhouse cucumbers usually begins at the blossom end of immature fruit.

Lesions are firm and become dark brown to black when cut open. Occasionally, lesions develop on one side of a fruit, causing it to hook as it grows.

Causal Organism

Gummy stem blight is caused by the fungus Didymella bryoniae. The pathogen can be seed-borne or can survive on organic debris from previously infected cucurbits or on wild or volunteer cucurbits. The gummy stem blight fungus produces two types of spores.

Windblown ascospores are likely to start the disease in a field. Later, pycnidiospores are released in a gummy substance that makes them adaptable for short distance spread by splashing water.

Spore production and infection are influenced by moisture and temperature. A moisture film from dew, rain, or overhead irrigation is necessary for spore germination. Optimum temperature for infection is 61 to 75 degrees F. Low night temperatures, particularly in greenhouses, may cause water droplets to exude from leaf points and condensation to form on leaves, favouring infection by the fungus at those points. Infection of fruits commonly occurs through harvest wounds or through dying flowers.

- Use only disease-free seed produced in arid western locations.
- Plow crop refuse deeply to reduce survival of the fungus as soon as crop is harvested.
- Practice crop rotation with non-cucurbit crops so that cucurbits are grown only every 3 to 4 years.
- Apply foliar-protectant fungicides on a routine basis.
- Avoid wounding fruit during harvest and store fruit at 45 to 50 degrees F to prevent postharvest fruit rots.
- Resistant cultivars are not currently available.

MUMMY BERRY OF BLUEBERRY

Mummy berry is one of the most serious diseases of blueberry. Once the disease becomes established in a planting, it can destroy most of the crop. Losses result from:

- Rotted berries;
- Killing or blighting of blossoms, blossom and leaf clusters and young shoots.

SYMPTOMS

Blighting of new shoot tips and blossoms can be easily mistaken for frost damage.

By blossom time, the infected young leaf and shoot growth will wilt, turn brown and die. About a week or so after infection of the early new growth occurs, dead areas develop on the petioles and along the midrib and veins of the leaves or at the base of flowers. Berries that develop from infected flowers may attain nearly full size before turning tan or gray and shriveling into hard mummies, which drop to the ground at or before harvest.

Causal Organism

Mummy berry is caused by the fungus, Monilinia vaccinii-corymbosi. The fungus overwinters in the shriveled mummies on the ground. In early spring, cup- or globe-shaped structures of the fungus called apothecia are produced on mummified berries during cool rainy periods. Spores produced inside apothecia are released into the air and carried by the wind to young developing leaf shoots and flowers where they cause primary infections. If moisture is not present, the fungus will not produce spores.

However, the fungus may survive in mummies for one year or more. Another type of fungus spores (conidia) is produced on dead tissue that results from primary infections. These conidia are spread during bloom by wind and insects, and result in secondary infection of flowers. Fruit that develops from infected flowers turns into mummies and falls to the ground.

Control

Removing mummified berries from the planting will greatly aid in controlling the disease. Removing these berries is not practical on a commercial scale, but may be of value in backyard plantings where there are just a few plants. After removing mummies, burn or bury them.

Cultivation in early spring to disturb or cover the mummies has been reported to be effective. Mummies that are disturbed or covered with soil at this time remain dormant or do not produce spores.

Cultivation between rows and raking under plants to disturb or cover mummies should be done as early as possible in the spring and repeated after each hard rain until after bloom. If just a few mummies are missed, they can produce enough spores to infect the planting. Where mummy berry is a problem, a good fungicide spray programme is essential.

LEATHER ROT OF STRAWBERRY

Leather rot of strawberry has been reported in many regions of the United States. In many areas, it is considered a minor disease of little economic importance. However, excessive rainfall during May, June, and July can lead to severe losses in fruit yield and quality resulting from leather rot. Commercial growers have lost up to 50 per cent of their crop to leather rot. The leather rot fungus primarily attacks the fruit, but many also infect blossoms.

SYMPTOMS

Leather rot can infect berries at any stage of development. Where the disease has been a problem, infection of green fruit is common. On green berries, diseased areas may be dark brown or natural green outlined by a brown margin. As the rot spreads, the entire berry becomes brown, maintains a rough texture, and appears leathery. The disease is more difficult to detect on ripe fruit. On fully mature berries, infection may result in little colour change or discolouration ranging from brown to dark purple.

Infected ripe fruit are usually softer to the touch than healthy fruit. When diseased berries are cut crosswise, a marked darkening of the water-conducting system to each seed can be observed. In later stages of decay, mature fruits also become tough and leathery. Occasionally, a white moldy growth can be observed on the surface of infected fruit. In time, infected fruit dry up to form stiff, shriveled mummies. Berries that are affected by leather rot have a distinctive, unpleasant odor and taste. Even the healthy tissue on a slightly rotted berry is bitter.

This presents a special problem to growers in pick-our-own operations. An infected mature berry with little colour change may appear normal and be picked and processed with healthy berries. Consumers have complained of bitter tasting jam and jelly made with berries from fields where leather rot was a problem. Leather rot is observed most commonly in poorly drained areas where there is or has been free standing water, or on berries in direct contact with the soil.

Causal Organism

Leather rot is caused by the fungus, Phytophthora cactorum. The fungus survives the winter as thick-walled, resting spores, called oospores, that form within infected fruit as they mummify. These oospores can remain viable in soil for long periods of time. In the spring, oospores germinate in the presence of free water to produce a structure called a sporangium.

A second type of spore, called a zoospore, is produced inside the sporangium. Up to 50 zoospores may be produced inside one sporangium. The zoospores have tiny flagella and can swim in a film of water. In the presence of free water on the fruit surface, the zoospores germinate forming a germ tube and infect the fruit.

In later stages of infection, sporangia are produced on the surface of infected fruit under moist conditions. The fungus is spread by splashing or windblown water from rain or overhead irrigation. Sporangia and/or zoospores are carried in water from the surface of infected fruit to healthy fruit where new infections occur.

Under the proper environmental conditions, the disease can spread very quickly. A wetness period (free water on fruit surface) of one hour is sufficient for infection. The optimum temperatures for infection are between 62 and 77 degrees F (17-25 degrees C).

As the length of the wetness period increases, the temperature range at which infection can occur becomes much broader. As infected fruit dry up and mummify, they fall to the ground and lie at or slightly below the soil surface. Oospores formed within the mummified fruit enable the fungus to survive the winter and cause new infections the following year, thus completing the disease cycle.

Control

- Select a planting site with good soil drainage and air circulation. Good soil drainage is critical. Sites that drain poorly or are subject to periodic flooding are ideal for leather rot development.

- Plants should be exposed to direct sunlight (avoid shade). Plant rows with the direction of the prevailing wind to promote faster drying of foliage and fruit.
- Mulch strawberry plants with straw or other material that reduces fruit contact with soil. Research has shown that a good layer of straw mulch is very beneficial in controlling leather rot.
- Proper spacing of plants and timing of fertilizer applications are important. Excessive applications of nitrogen fertilizer can produce excessive amounts of dense foliage. Shading of berries by thick foliage prevents rapid drying of fruit during wet periods and creates ideal conditions for disease development.
- Pick fruit frequently and early in the day (as soon as plants are dry). Cull out all diseased berries, but do not leave them in the field.

CANE BLIGHT OF RASPBERRIES

Cane blight is one of the more damaging diseases of raspberries. The disease is most common on black raspberries but also occurs on red and purple varieties. The disease occasionally occurs on blackberries and dewberries. Cane blight can result in wilt and death of lateral shoots, a general weakening of the cane, and reduced yield. It is usually most severe during wet growing seasons.

SYMPTOMS

On first year canes (primocanes) dark brown-to-purplish cankers form on new canes near the end of the season where pruning, insect, and other wounds are present. The cankers enlarge and extend down the cane or encircle it, causing lateral shoots above the diseased area to wilt and eventually die. Black specks, which are reproductive bodies of the cane blight fungus, develop in the brown cankered bark. In wet weather, large numbers of microscopic spores ooze out of the pycnidia.

This ooze gives the bark a dark-gray, smudgy appearance. During winter, infected canes commonly become cracked, brittle, and snap off easily. On infected second-year canes (floricanes), the side branches may suddenly wilt and die, usually between blossoming and fruit ripening.

Upon close examination, the presence of dark brown or purplish cankers can be observed on the main cane or branches below the wilted area.

Causal Organism

Cane blight is caused by the fungus, Leptospaeria coniothyrium. The pathogen survives over winter on infected or dead canes. The following spring, spores are released and carried by splashing rain and wind to nearby primocanes. Under moist conditions, the spores germinate and penetrate pruning wounds, insect punctures, fruit stem breaks and other wounds.

After entry the fungus rapidly invades and kills bark and other cane tissues. Fungal fruiting bodies are formed in older cankers and complete the disease cycle. Dead canes continue to produce conidia and remain a source of infection for several years.

Control

- All steps possible should be taken to improve air circulation within a planting, to allow faster drying of foliage and canes. Reducing the number and duration of wet periods should reduce the potential for infection. Excessive applications of fertilizer (especially nitrogen) should be avoided, since it promotes excessive growth of very susceptible succulent plant tissue.

  Plants should be maintained in narrow rows and thinned to improve air circulation and allow better light penetration. Weeds are very effective in reducing air movement; therefore, good weed control within and between rows is important for improving air circulation within the planting. Raspberries should be planted in sunny, open areas where water and air drainage are good. This allows plants to dry quicker after wet periods, and reduces the chance of infection.

- Wild brambles, especially wild raspberries, growing in the area should be removed. They can provide a continuous source of spores to spread this and other diseases and pests to cultivated raspberries and blackberries.
- Healthy, rapidly growing plants that have been properly fertilized and watered are more resistant to cane blight.
- After harvest, remove and destroy all old fruited floricanes and any new primocanes canes that are infected. Old canes should be removed before growth starts in the spring.
- Keep plantings free of insects, since they may cause wounds that serve as entry points for the fungus. Avoid any other pests or cultural practices that result in wounding of the canes.
- If cane blight is a serious problem, the use of fungicides should be considered.

EUTYPA DIEBACK OF GRAPE

For more than 60 years the eastern grape industry recognized a disease called "dead-arm" thought to be caused by the fungus Phomopsis viticola. In 1976, researchers demonstrated that the "dead-arm" disease was actually two different diseases that often occur simultaneously: Eutypa dieback and Phomopsis cane and leaf spot.

Eutypa dieback is the new name for the trunk and arm phase of what was once known as "dead-arm."

Scientists now propose that the name "dead-arm" be dropped. It is important that Eutypa dieback and Phomopsis cane and leaf spot be considered as two distinctly different diseases because the control recommendations for each are quite different.

SYMPTOMS

The earliest symptom is a canker that generally forms

around pruning wounds in older wood of the main trunk. These cankers are usually difficult to see because they are covered with bark. One indication of a canker is a flattened area on the trunk. Removal of bark over the canker reveals a sharply defined region of darkened or discoloured wood. These cankers may be up to 3 feet long, but do not enter one- or two-year wood and seldom go below ground line.

When the trunk is cut in cross-section, the canker appears as darkened or discoloured wood extending in a wedge shape to the centre of the trunk.

The most striking and obvious symptoms of Eutypa dieback are the leaf and shoot symptoms, which may not develop for up to 3 years after infection of the vine has occurred. These symptoms are most obvious in spring (May and June). Spring shoot growth is weak and stunted above the cankered area. Leaves are at first smaller than normal, cupped, misshapen and yellowed.

Later in the season (mid July), these leaf and shoot symptoms may disappear from all but the basal leaves of affected shoots. The vines may appear to have recovered. However, the infected trunk and all growth above it will eventually die.

Causal Organism

Eutypa dieback is caused by the fungus, Eutypa armeniacea. The fungus survives in infected trunks for long periods of time, whether they remain as part of the in-place vine or as prunings in the vineyard. Eventually, the fungus produces reproductive structures (perithelia) on the surface of infected wood. Spores (ascospore) are produced in these structures and are discharged into the air. Ascospore discharge is initiated by the presence of free water (either rainfall or snow melt).

Most spores appear to be released during winter or early spring, with relatively few being released during the summer months. Unfortunately, most spores are released at the same time pruning is being conducted. The ascospore can be carried considerable distances by air currents to recent wounds on the trunk. Pruning wounds are by far the most important points of infection. When the ascospore come in contact with newly cut wood, they germinate and a new infection is initiated.

Control

- At present the primary control methods is removing infected trunks from the vineyard. The vine must be cut off below the cankered or discoloured wood. If the canker extends below the soil line, the stump can be left and a new trunk formed. It is important to remember that the best time to identify infected vines is spring (May and June) and when the leaf and shoot symptoms are most obvious. If trunks cannot be removed in the spring, they should at least be marked so they can be removed after harvest, but before the next spring.
- Sanitation is critical. All wood from infected plants must be removed from the vineyard and destroyed (either buried or burned). An old infected trunk lying on the ground may continue to produce spores for several years.
- At present, no fungicide recommendations are available for control of this disease.

NECROTIC LEAF BLOTCH OF GOLDEN DELICIOUS APPLES

Necrotic leaf blotch is a common but minor disease that appears to be restricted to the Golden Delicious cultivar of apples and its sports. The disease is a physiological dis-order and has been observed for about 20 years in most apple-growing areas of the Eastern and Midwestern United States. Necrotic leaf blotch usually appears in late June or early July and is most severe in the latter part of the growing season.

Mature leaves from the base to the centre of upright, succulent, rapid-growing shoots are usually the only ones that show symptoms. Young, immature leaves on succulent shoots and mature or cluster leaves on fruiting wood are usually not affected. The effect of the disease varies among individual Golden Delicious trees, both from orchard to orchard and from tree to tree within the same orchard.

SYMPTOMS

The symptoms are a leaf blotching, followed by a yellowing of the affected leaves (chlorosis) and defoliation. Defoliation can be severe. Irregular brown blotches, usually 0.5 to 1. 5 centimeters in diameter, suddenly appear on leaves. The blotches are usually restricted by the larger leaf veins. Many affected leaves turn yellow and drop all at once a few days after the necrotic blotches first appear.

The most conspicuous symptom of necrotic leaf blotch is the large number of yellow leaves, on the tree and later on the ground, that suddenly appear during July and August. A distinctive characteristic of the disease is its appearance in two to four, or more, waves of these symptoms during certain periods that are scattered throughout the latter half of the growing season. Between these periods or waves, little or no additional disease develops. During the course of the season, 10 to 50 per cent defoliation may occur on severely affected trees. All sports of Golden Delicious on any root stock appear to be equally susceptible.

Cause

Necrotic leaf blotch of Golden Delicious apples appears not to be caused by a fungus, bacterium, air pollutant, or a nutrient deficiency. The causal agent of necrotic leaf blotch is unknown; therefore, the disease is often referred to as a physiological disorder. The disease commonly appears during hot, hazy weather following a humid, rainy period. Other environmental factors or an air pollutant may be involved. The levels of ozone or sulfur dioxide in the air and foliar nutrients in the leaves have little or no effect on disease development.

Control

- It has been observed that certain fungicides result in

less necrotic leaf blotch. Since a fungal pathogen does not appear to cause necrotic leaf blotch, it is difficult to assess the role of fungicides in reducing necrotic leaf blotch. The effect may be indirect, by altering leaf physiology.

Fungicides containing zinc (Zn) ions appear to be the most effective. Golden Delicious trees sprayed, during the summer cover spray period, with fungicides containing mancozeb (Dithane M-45, Manzate D, Penncozeb) or Ziram tend to have less necrotic leaf blotch than unsprayed trees or trees sprayed with other fungicides.

- Trees that have an annual moderate crop of fruit have less necrotic leaf blotch than trees that have a biennial bearing habit or a light fruit crop.
- Golden Delicious trees pruned to a central leader commonly have less necrotic leaf blotch than trees pruned to an open centre.

WHITE PINE BLISTER RUST ON CURRANTS AND GOOSEBERRIES

White pine blister rust is not a serious disease of currants and gooseberries; however, it is a very serious disease of white pines (Pinus strobus). Currants and gooseberries serve as an alternate host for the rust fungus that causes white pine blister rust. Therefore, planting currants and gooseberries in areas where white pines are present can lead to serious losses of white pines. North American white pine species, including bristlecone, limber, sugar, eastern white, southwestern white, western white, and whitebark, are highly susceptible. White pine blister rust causes significant damage in pine forests by forming cankers on the branches of white pines.

These cankers ultimately kill the trees. Black currant is the most susceptible of the Ribes species. To protect white pine forests, several states have enacted laws concerning planting of black currants. The current law to suppress and control White Pine Blister Rust Disease is as follows:

- The European black currant, Ribes nigrum L. or any

variety of this species is hereby declared to be a public nuisance, and it shall be unlawful for any person to possess, transport, plant, propagate, sell, or offer for sale, plants, roots, scions, seeds, or cuttings of these plants in this state.

- Recognized varieties, e.g., "Consort" produced by the hybridization of Ribes nigrum L. or a variety thereof with a resistant or immune species, known to be immune or highly resistant to the White Pine Blister Rust fungus.

SYMPTOMS

On Ribes in the spring, tiny yellowish spots become visible on the upper surface of the leaves, while on the underside, orange-yellow blister-like fruiting bodies appear. By late summer, yellow to brown threadlike growths develop on or near these infection spots on the leaf. Bushes also will have premature defoliation.

On white pine, the symptoms include dead branches, chlorotic foliage, branch girdling by lesions that exude resin or sticky yellowish fluid (spermagonia), cankers that are diamond-shaped to elliptic with a dead centre surrounded by a band of yellowish-green infected bark, light yellow-orange aecia, and death of the tree.

Causal Organism and Disease Development

White pine blister rust is caused by a fungus, Cronartium ribicola. The organism was introduced from Europe in the early 20th Century. It has spread throughout the entire range of white pines in North America. The life cycle takes three to six years to complete. The initial infection of black currant bushes occurs in the spring, when aeciospores from diseased white pine land on the leaves of the bush.

These spores can travel on the wind several miles. Moisture is needed for the germination of the aeciospores. After one to three weeks incubation, the plants begin to show the first symptoms of disease. Yellowish spots appear on the top side of the leaves and fruiting bodies (uredia) appear on the underside of the leaves. The fruiting bodies (uredia) produce uredospores which reinfect Ribes leaves. This cycle can be repeated many times during a single growing season. In late summer or early fall, when day length and temperature decrease, telia (another type of fruiting body) replace the uredia on the underside of the leaves.

The telia produce teliospores, which germinate in place at night during wet weather and produce basidiospores. After telial formation, at least 60 hours of wet weather with temperatures not exceeding 20 degree C are required for basidiospore formation, dispersal, and infection of white pine. These basidiospores are carried by the air currents to white pine trees. The basidiospores germinate on wet needles and enter through the stomata.

The only symptom initially is a yellow to reddish spot at the site of infection. Infected needles turn yellow and drop prematurely, but often not before the fungus has grown down the needle and entered the twig.

This occurs late in the fall or early in the spring of the next year. About 12 to 18 months after infection, the fungus has grown from the needle into the bark of the stem, where it produces spermagonia.

During the first or second year of infection, the spermagonia produce aecia. The aecia produce powdery bright yellow aecidiospores that may live for many months. The aecidiospores are carried by the wind to Ribes bushes, completing the cycle.

Chapter 5

Impact on Chocolate Production

Chocolate has its origins in ancient Central America where the Maya and the Aztecs cultivated the cacao tree (*Theobroma cacao* L.) and extracted from the seeds or beans a highly prized drink, which was called *chocolatl,* a precursor to our modern English word chocolate. *Theobroma* means the source of the "food of the gods," hence its scientific name Theo (god) and broma (food).

The Spaniards, notably Hernan Cortez, introduced the drink to the Spanish royalty, but the bitter drink did not become popular for another one hundred years, when additives such as sugar, cinnamon, and chile peppers made the drink more palatable. Chocolate was the first mildly stimulating drink introduced to Europe, appearing even before coffee and tea.

The present-day popularity of chocolate and the drink cocoa needs no verification as attested to by the many "chocolate lovers" who especially look forward to occasions and national holidays when chocolate is traditionally given as a gift to loved ones and friends.

The demand for chocolate is growing and the question now is will the world supply of a product that comes from a strictly tropical, rainforest-inhabiting tree will continue to meet the demand.

Over the past two decades several important fungal diseases have gained considerable importance and pose a serious threat to the supply of chocolate.

Depending on where cacao is grown, one or more of three

diseases (black pod, witches' broom, and frosty pod rot) may reach epiphytotic proportions that cause devastating losses.

Table: Estimated Annual Reduction in Potential Cocoa Production by Major Diseases.

Diseases	*Pathogen*	*Region*	*Reduced Production*	
			(tons ×1000)	*($ million)**
Black Pod	*Phytophthora* spp.	Africa/Brazil/ Asia	450	423
Witches' Broom	*Crinipellis perniciosa*	Latin America	250	235
Frosty Pod Rot	*Moniliophthora roreri*	Latin America	30	47
Swollen Shoot	CSSV	Africa	50	28
Vascular-streak dieback	*Oncobasidium theobromae*	Asia	30	28

In Brazil, the production of cocoa beans has dropped from 400,000 to 100,000 metric tons in just 10 years, largely as the result of the infection by the fungus, *Crinipellis perniciosa*. This fungus has spread beyond Brazil into Peru, Ecuador, Venezuela, and Colombia in South America, and Panama in Central America. The pathogen also is on the Caribbean islands of Trinidad and Tobago.

A second fungus, *Moniliophthora roreri*, causes another very damaging disease, frosty pod rot, that also destroys the cacao beans inside the pods. This pathogen has not yet moved into Brazil, but has moved beyond Panama in Central America into Costa Rica and Nicaragua. Vast numbers of air-borne spores of the fungus, that give the pods a frosty appearance, will undoubtedly spread the disease to other cacao growing regions.

The disease that causes the most widespread destruction of cacao worldwide is black pod, which is caused by several species of *Phytophthora*, a fungus-like microorganism that comes in several forms. The most widespread species are found around the world in all cacao growing areas, and cause sporatic losses. One species, *P. megakarya*, is spreading throughout West Africa, the world's premier cacao growing area, and causes major losses.

Vascular streak dieback, caused by *Oncobasidium theobromae*, is found only in Asia, and swollen shoot virus disease, found only in West Africa, are of relatively minor importance compared to the other above-mentioned diseases.

Many factors contribute to a decline in production of cocoa beans worldwide, including insect infestations, social pressures to grow other crops, economic issues that discourage long-term commitment of small acreage farmers to grow the crop, and societal pressures to destroy rain forest environments. Plant diseases are major components of the decline in production. We as plant pathologists and microbiologists have as our mandate to discover and devise means to reduce disease losses and to save chocolate for the enthusiastic consumers of the world.

BLACK POD

Black pod disease of cacao is an economically serious problem in all areas of the world where cacao is grown. Annual losses due to black pod may range from 30-90% of the crop. This disease is especially severe in West and Central Africa which contributes 60-70% of the world production of cocoa beans. The disease is caused by a complex of species of *Phytophthora* (the genus named by Anton de Bary in 1876 as "plant destroyer"). These pathogens are fungus-like microorganisms with a world-wide distribution. Species of *Phytophthora* are pathogens on most economically important crops grown throughout the world, such as *P. infestans*, the cause of late blight of potato.

Previously, isolates of *Phytophthora* from cacao were classified as *P. palmivora* (Butl.) Butler. This species has been recognized as one of the most important pathogens in the tropics attacking many plantation crops, including cacao, rubber, black pepper, coconut, pineapple, and papaya, as well as citrus, avocado, and many ornamental plants. Early studies of the pathogen revealed some variation in the morphology and types of lesions produced on cacao pods among isolates from different countries. In 1976, isolates of *P. palmivora* from cacao were placed into one of four morphological groups (MF1, MF2, MF3, and MF4). Further studies redefined the morphological groups. The MF1 form was considered to be the typical *P. palmivora*, and the MF2 form a variant (atypical) of *P. palmivora*. The MF3 form was described as a new species, *P. megakarya* Brasier and Griffin, and the MF4 form considered to be *P. capsici* Leonian.

Recently, it has been proposed that isolates of *P. capsici* differing in morphological and pathological attributes should be called *P. tropicalis*. *Phytophthora palmivora* is present in most countries and is an important part of the black pod complex, while *P. megakarya* is present only in several countries in West Africa. However, *P. megakarya* appears to be more virulent than *P. palmivora* and is becoming the dominant species in West Africa, moving from Nigeria and Cameroon into countries where it has not been reported previously.

In the major cacao-growing region of Bahia, Brazil, three species have been implicated: *P. palmivora*, *P. capsici*, and *P. citrophthora* (Smith and Smith) Leonian. *Phytophthora capsici* appears to be the dominant and most important species attacking cacao in Brazil, and also has been reported to occur in other countries in Central and South America, as well as in the West Indies, Indonesia, and India. *Phytophthora citrophthora* is the least common species found in Brazil, but is the most virulent. *Phytophthora heveae* Thompson also has been found to cause black pod in some countries.

Although *Phytophthora* species attacks all parts of the cacao plant, the major economic loss is from infection of the pod. Pods or cherelles (immature pods) may be infected at any place on the surface, but infection is most often initiated at the tip or stem end.

The disease causes a firm, spreading, chocolate-brown lesion that eventually covers the whole pod. The beans inside the pod may remain undamaged for several days after initial infection of the husk, thus frequent harvests may prevent much yield loss. In advanced infections, *Phytophthora* invades the internal pod tissues and causes discoloration and shriveling of the cocoa beans. Diseased pods eventually become black and mummify. The pathogen also will cause a seedling blight in cacao nurseries as a result of infection of the stem and young leaves.

The inoculum that initiates black pod can come from the soil and/or infected roots, stems, and leaves. Root infection from residual soil inoculum usually is not an economic concern, however, the pathogen is capable of producing spores on all infected plant parts.

Thus, infected roots may serve as a source of inoculum for infection of the pods. Bark and stem cankers may function in the same manner. Once a pod is infected and sporulating, it may then provide a massive source of inoculum to infect other pods.

Under humid conditions, sporangia (asexual reproductive structures and a source of secondary inoculum) may form on the surface of infected roots, cankers, or diseased pods. This type of propagule can be dispersed by rainfall, splashing water, and water moving over the surface of the soil. When sporangia are in free water, zoospores are formed inside the sporangia and released into the water. Zoospores are motile spores that actively swim towards an infection site and are the primary infective propagules. Thus, windborne rain is a primary factor in the spread of the disease. Ants also have been reported to transport inoculum.

Control of black pod is difficult because *Phytophthora* can persist in soil and debris for several years. Also, since susceptible pods may be present on the trees most of the year, the pathogen may always be present in the canopy, ready to cause major epidemics when environmental conditions become favorable for sporulation and dispersal. Frequent harvesting will lessen the danger of spread of the disease from infected pods.

Regular pruning to remove infected chupons (small suckers at the base of the tree) and increase air circulation (to reduce the humidity under the canopy) is an important disease management tool. Other measures, such as the removal of infected pods and husk piles, may have some effect on inoculum levels. However, under high rainfall conditions, it is difficult to effectively suppress the inoculum to attain good control. Chemical control relies on the use of copper and metalaxyl-based fungicides. However, these are not entirely effective, are expensive for the small farmer, and not economically feasible.

Additionally, there are environmental concerns regarding the heavy use of chemicals in the tropical rainforests, and there may be problems with non-target effects and resistance of the pathogen. At present, there is no acceptable genetic resistance in cacao to control black pod.

Active research is underway throughout the world on this important disease of cacao. Control measures being investigated are aimed at integrated pest management (IPM) strategies, and include biological control microorganisms, genetic and induced resistance, cultural practices, natural products, and limited use of chemicals.

WITCHES' BROOM

Witches' broom disease, caused by the fungus *Crinipellis perniciosa* (Stahel) Singer (formerly *Marasmius perniciosus*), was first reported in 1895 from Surinam and is currently found in several countries of South America and the Caribbean islands. The disease is initiated by basidiospores produced and released from pinkish mushrooms called basidiocarps, that are dispersed by wind and rain onto leaves, flowers, and fruit (pods) of cacao plants.

It is estimated that a single basidiocarp can release 80 to 90 million basidiospores. In the presence of free moisture (rain and dew) and high relative humidity, basidiospores germinate and penetrate young meristematic tissues in vegetative and floral buds through stomata, epidermis, or trichomes.

The colonized tissues undergo several physiological and hormonal changes leading to swelling and formation of numerous succulent vegetative branches, known as brooms, within flower cushions and on vegetative apical or axillary buds. The brooms are usually formed within 5 to 6 weeks following infection. The fungus also infects pods causing necrotic lesions, uneven ripening, and various deformations. Brooms and infected pods become progressively brown and dry, and within 3 to 8 months following broom and pod drying, basidiocarps are formed following several alternating wet and dry periods.

Witches' broom debilitates cacao trees by diverting energy to the numerous brooms that do not form flowers, thus reducing yield potential. Further yield loss is incurred by the prevention of seed formation in pods infected early in their development. If seeds are formed prior to infection they may be unusable depending on the extent of pod colonization by *C. perniciosa.*

In its natural habitat of the Amazonian forest, cacao is found as scattered under story trees which are genetically variable, a characteristic that hampers the development of epidemics of witches' broom. The cultivation of cacao has intensified over the years as cocoa has progressively become a global trade commodity. Production environments that encourage the planting of cacao varieties with little genetic diversity ensures a constant supply of susceptible tissue for infection by *C. perniciosa.*

Management of witches' broom has received considerable attention from the beginning of the 20th century. Broadly, there are four major strategies that may be adopted: phytosanitation, chemical control, genetic resistance, and biological control. Phytosanitation, by removal and destruction of diseased plant parts, has been shown to reduce pod loss and delay disease epidemics.

However, this strategy is tedious and in one study it has been shown that 95% removal is required to achieve 50% reduction in pod loss. Chemical control of witches' broom with protectant and systemic fungicides is not a routine practice in cocoa production because of high costs and risks associated with cocoa bean contamination and environmental health.

The development of genetically resistant cacao cultivars is an on-going endeavor in many countries, and it is expected that the use of these cultivars would reduce the incidence of the disease. Prospects of managing witches' broom disease of cacao through biological control have been investigated for over 20 years leading to the isolation of a new species, *Trichoderma stromaticum*, a parasite on the mycelium and basidiocarps of *C. perniciosa*. In Brazil, commercial formulations of *T. stromaticum* are currently used in managing witches' broom.

However, the inconsistent performance of *T. stromaticum* indicates the need to further understand the relationship between the cacao plantation environment and the survival and establishment of this mycoparasitic fungus. Collaborative research efforts by scientists at research institutions in the United States and Central and South America are currently underway to optimize the use of *T. stromaticum*.

FROSTY POD ROT

Frosty pod rot (or Moniliophthora pod rot), caused by *Moniliophthora roreri* Evans et al., originally described in 1933 as *Monilia roreri*, is a devastating disease of cacao pods. The conidia (the only known infective propagules) infect by penetrating the surface of the pods. The pods are highly susceptible during the first 90 days of their growth. Early symptoms include discolored areas of swelling on the pods followed by a dense formation of cream-colored spores, the so called frosty pod, developing on the pod surface within 2 weeks after infection. The area of the pod with sporulation spreads rapidly, and spore densities have been estimated to be as high as 44 million conidia per square centimeter. The spores are later released by wind or by water droplets during rainy periods.

Frosty pod rot occurs on all *Theobroma* and related *Herrania* species. The first report of the disease was by J. B. Rorer while on a scientific trip to Ecuador to investigate serious losses in yields due to diseases. Frosty pod rot currently is restricted to northwest South America (Ecuador, Peru, Colombia) and south Central America (Nicaragua, Costa Rica and Panama). However, it is an immediate threat to Brazil, which is one of the largest cacao growing areas in South America. Yield losses have been estimated between 25% to total loss of the crop if frequent (7-10 day schedules) phytosanitary measures are not implemented. The most economical way of controlling the spread of this disease is to eliminate the inoculum sources, which are dead sporulating pods, with frequent harvests on regular cycles. Other practices should include reduction of tree height (maximum 3.5 m) to facilitate removal of diseased pods, biological control, and planting of resistant or tolerant cultivars as they become available.

In view of the difficulty in finding a practical and cost-effective fungicidal treatment, alternative control strategies, such as biological control are being investigated in Peru, Panama, and Costa Rica to manage frosty pod rot. This involves the use of naturally occurring fungal agents (mycoparasites) capable of parasitizing the pathogen and preventing the dispersal of fungal spores. Saprophytic fungi belonging to the genus *Trichoderma* and *Clonostachys*, selected for their parasitizing capacity, are being tested in field trials. Bi-weekly to monthly sprays of biocontrol agents have been shown to reduce the production of conidia on diseased pods. Biocontrol treatment along with phytosanitation has reduced pod loss and increased yields in previously abandoned cacao plantations.

Major genetic improvement efforts are underway in Costa Rica, Colombia, and Ecuador to select genotypes resistant or tolerant to *M. roreri*, and new hybrid clones are being distributed to farmers. Breeding for disease avoidance may be one of the safest methods to reduce field losses, as this form of resistance is less vulnerable to adaptive changes by the pathogens. For example in Ecuador, disease losses due to *M. roreri* were found to be lower in trees that bring a high proportion of their total pods to maturity during the later months of the dry seasons or early wet season, thus avoiding conditions that favour disease development.

GENETIC AND INDUCED RESISTANCE IN CACAO

Cacao is an open pollinated (cross breeding) species. This poses difficulties when trying to improve such characteristics as bean quality and yield, or disease and insect resistance in cacao. Individual trees bearing beneficial traits fail to produce offspring that consistently bear the same traits. When farmers select seed from their best trees, the resulting trees are often of poor quality, produce low yields and are susceptible to diseases and insects. The only way to produce trees of uniform high quality is by vegetative propagation resulting in exact clones of the parent tree.

A number of vegetative propagation methods have been developed. These include low-tech rooted cuttings and grafting techniques as well as high-tech tissue culture and micropropagation techniques.

Unfortunately, high-tech solutions are expensive, requiring government or industry support, and low-tech solutions are slow. Considering the intensive pressure from diseases and insects on the world cacao crop, these techniques fail to meet the farmers' needs for high quality seedlings. A combination of low-tech and high-tech methods has been worked out and offers hope for the production of large numbers of high quality seedlings in the future.

There is great potential for accelerating the improvement of cacao using traditional plant breeding enhanced by molecular genetics techniques.

These techniques allow breeders to maintain and combine important agronomic traits controlled by many genes, such as yield and disease resistance, in their breeding populations, and produce new varieties for release with fewer generations to farmers. Breeders often have limited access to the best germplasm sources. Some of the sources may have yet to be discovered in wild populations in South and Central American forest or may be located in collections separated by great distances and national boundaries.

Through international collaborations, scientists using molecular techniques are attempting to catalogue and characterize the genetic diversity in the world's cacao populations, including both wild and cultivated populations. It is estimated that there are 18,000 cacao accessions in collections worldwide, and these accessions can be grouped into 2500 to 3000 genetic groups. An additional 300 to 500 genetic groups may exist in wild populations. As more is learned about the world's cacao populations, we should be better able to exploit this valuable resource to improve not only production of chocolate, but the lives of many small farmers dependent on cacao production worldwide.

The combination of modern plant breeding and plant propagation techniques should allow these improvements to have an impact on global cacao production more rapidly than ever thought possible in the past.

The genetic control of disease resistance traits can be simple, involving a single gene, or complex, involving many genes. Resistance to disease in cacao has been identified in some cases, but limitations to increasing the number of trees carrying those traits prevent their wide spread use. In addition, pathogen populations differ between cacao production areas, so resistance that is effective in one area may totally fail to provide disease control in another area. Novel disease control methods such as biological control using beneficial microorganisms, in some cases, can function regardless of the genetic background.

Biocontrol offers a potential control strategy for the cacao trees presently under cultivation. Most, if not all, plant species possess the ability to resist disease development. Disease develops when the plant's defence mechanisms are not activated. This phenomenon has provided the rationale behind the development of a new class of disease control compounds that activate plant defence mechanisms allowing an otherwise susceptible plant to protect itself. These types of compounds are now being tested for disease management in cacao.

Some of the same factors that limit the potential for using traditional chemical control measures limit the use of biocontrol. Cacao is often grown in remote areas and as an understory tree.

This situation, combined with the low market value for the product and the high cost of supplies and labour, often make any disease control strategy requiring regularly scheduled applications financially impractical. Perhaps an optimum disease management strategy would be to identify beneficial microorganisms that persist in the cacao canopy and root systems and provide disease control without regular spray applications.

These types of organisms, epiphytes, endophytes, and mycorrhizae exist in cacao and are currently being studied for their beneficial effects. It is possible that microorganisms that induce resistance to disease and insects in cacao can be identified. The use of these microorganisms could result in long term disease control with limited cost to the farmer and could actually be beneficial to the environment.

SUSTAINABLE CACAO PRODUCTION

The chocolate industry is keenly aware of the requirement for an adequate supply of cocoa beans of consistent quality. Therefore, it is essential that the industry and all its associated institutions and interests support scientific research on cacao and its products in key areas such as production, integrated pest and disease management, germplasm conservation and improvement, and biotechnology.

The cacao research community is international in scope. Industry, trade organizations, government, and non-government organizations support the comprehensive effort to address ecological, social, pest and disease management, economic, and cacao improvement concerns.

The USDA/ARS alone has formal and informal research agreements with many organizations, including those listed in Table, and with other government and nongovernment organizations, and universities.

Cultivation of cacao within a sustainable, biologically diverse agricultural system also provides habitat to important forest and migratory bird and mammal species.

In Central and South America and in Africa, strategic preservation of rain forest remnants and the development of forest links have offered hope to efforts to preserve rain forest environments.

Sustainable cacao will not only help the cocoa and chocolate manufacturers maintain a constant and reliable supply of raw material, but it also will aid in the global efforts to conserve tropical forest ecosystems and move small holder farmers into a more favorable economy.

Table: Organizations with formal or informal research agreements with USDA/ARS.

Organization	*Country*
ACRI	USA
M&M MARS CORP.	USA
USAID	USA
CICAD	OAS
CABI-BIOSCIENCE	UK
SENASA	Peru
CEPLAC/CEPEC	Brazil
CORPOICA	Colombia
CATIE	Costa Rica
IRAD/IITA	Cameroon
Smithsonian Inst (STRI)	Panama
University of Maryland	USA
Penn State University	USA
D^2 Biotechnologies	USA

According to Young, prehistoric peoples in Mesoamerica had the correct approach to farming cacao. Relatively small cacao orchards enclosed by forest or set within small, diversified plots of various crops within the forest optimized the chances for high productivity. Pollination and containment of diseases and insects under these conditions may have been better than what is typically found in large, monoculture-type plantations today.

The challenges of disease epiphytotics, insect infestations, poor pollination, and chemical fertilization were minimized because of the practices of the Mayans, Aztecs, and other mesoamerican people. They knew that growing "the food of the gods" in small groves reaped considerable harvests.

Understanding the components of cacao crop management is essential to growing sustainable cacao. The impact of diseases such as black pod, witches' broom, and frosty pod rot will continue to be devastating until progress is made to control these diseases. Short term solutions such as chemical fungicides are often cost-prohibitive, may be damaging to the environment, and leave undesirable residues in the product. Long term solutions such as biotechnological approaches or breeding for disease resistance are time-consuming to develop and are not readily available for immediate application.

The most immediate answer to the disease problems is to use what tools are presently available in an integrated approach. These include limited biological control methods, disease resistant material, crop sanitation to remove diseased material as a source of inoculum, and cultural practices such as shade optimization, pruning, and soil fertility management.

As stated by Young, the cacao tree is truly a creature of nature and is cultivated in a relatively primitive way compared to modern row crop agriculture. It may be compelling to protect the tropical ecosystems that are still left to ensure the world's coveted supply of chocolate and at the same time preserve many more biological riches integrated into the complex fabric of tropical diversity.

Chapter 6

Economics of Plant Disease

Increasing trade liberalisation, globalisation and international transportation of people and commodities have increased the potential for disease incursion, both plant and animal, in countries like Australia. While a comparatively strict quarantine regime and geographic isolation provide substantial protection in Australia, disease incursions are not uncommon.

In recent years, there have been several exotic disease outbreaks including wheat stripe rust, bacterial blight of cotton, sugar cane ratoon stunt, potato cyst nematodes, karnal bunt, grapevine leaf rust, papaya fruit fly, Newcastle disease in poultry flocks, and Ovine Johne's Disease (OJD) in sheep. Recent attention on the incursion of plant diseases followed the outbreak of *black sigatoka*, a banana leaf disease, in the Tully district of North Queensland in 2001, and citrus canker, a highly contagious bacterial disease for citrus fruits, in the Central Highlands region of Queensland in 2004.

An outbreak of any pest or disease, either plant or animal, has the potential to be widespread in countries like Australia, with subsequent economic, social and environmental impacts. At the same time, the trade liberalisation process is focused on avoiding quarantine being used as a trade barrier, where the risks of disease incursion are used to 'trump' any potential benefits of increased trade flows. There is a strong economic case for an efficient quarantine regime, with import, monitoring and surveillance procedures, and for appropriate eradication measures when disease outbreaks occur. Estimates of the economic cost of outbreaks, and the benefits of avoiding them, are needed to help weigh policy options about appropriate future prevention and response strategies.

A number of studies have been conducted on the entomology and epidemiology of introduced diseases. However, little research attention has been paid to the economics of such outbreaks.

In Australia, there has been work in recent years in the area of weed management which is analogous to the disease problem in terms of containment options and timing.

Although public policy evaluation and administration of plant disease incursion issues are considered to be important, they remain understudied. The full extent of the costs of damages caused by pest and disease incursions has only recently received greater attention. There is only a limited number of studies attempting to quantify the economic losses of plant disease outbreaks and the desirability of appropriate response strategies.

To estimate the economic cost of an outbreak, proper identification of losses incurred and their quantification are important. Costs of an outbreak can be generated from three different sources:

- Surveillance, control and eradication;
- Losses of production, income and employment and its effects on ancillary industries;
- Loss of trade access. In addition to these costs, there might be non-market impacts and costs to consider.

For example, impacts on biodiversity will not be reflected in market transactions, and would require specialised nonmarket valuation techniques in the assessment stage. At a policy-making level, there is often interest in showing that the benefits of control measures outweigh the costs. There may also be interest in evaluating what the net benefits of different control measures should be.

Techniques of economic analysis can be used to estimate the economic cost of an outbreak. In this paper, an economic analysis of the government response to the 2004 citrus canker outbreak in Queensland is provided. The analysis is focused on two key issues. The first is to confirm whether control measures were appropriate, while the second is to identify what level of control would have been most effective. An additional consideration for this paper is the economic issues involved in potential compensation for growers affected by the outbreak.

The importance of economic analysis and provide a review of the techniques employed to estimate the economic impact of disease outbreaks. A comparative analysis of the national response strategy to the citrus canker against the alternative Pressler Plan. The penultimate section presents an economic analysis of optimising the response strategy on eradication and control measures.

THE IMPORTANCE OF ECONOMICS OF DISEASE INCURSION

An economic framework can be used to explain why a government response is typically needed for the prevention and control measures in a disease infected industry. Relying on control by private firms is usually ineffective because any incursion of a disease has an external effect on market activities. Costs are not borne solely by the source of the incursion, but spillover (intentionally or unintentionally) into other areas, without affected parties being able to exact a compensation for the damage incurred.

Therefore, many of the impacts of a disease outbreak are spillovers in the sense that the market mechanism is unable to transmit back the true cost of a plant disease outbreak to an infected source. For instance, neither commercial nor backyard scale citrus growers have adequate incentives for voluntary eradication.

Citrus canker containment and eradication actions demonstrate many of the characteristics of a public good. Disease-free industry status has aspects of non-rivalry in consumption (that is, one orchard's ability to enjoy disease-free status does not deprive others) and nonexclusion in provision (that is, it is difficult to exclude growers from enjoying this status).

Growers often face powerful incentives not to report disease incursions to the authority responsible for monitoring and surveillance, because the removal of trees can impose major financial losses to orchards. The first infected farm in

Emerald sought an injunction from the court to stop trees being cut and destroyed after the original outbreak in June 2004. In Florida, a legal battle continued for some time to determine whether the regulatory authority should have the right to trespass the property to test for canker infestation and to remove diseased trees. However, failure to control the disease would result in negative external effects on other growers as citrus canker would spread to other residences and commercial orchards.

Therefore, the costs to other growers are not usually internalised into private decisions, but are of concern to public regulatory agencies and other stakeholders. Furthermore, prevention of disease outbreaks requires extensive monitoring and surveillance both by the regulatory authority and individual grower.

On the part of the growers, this imposes additional costs on them. While there may be market incentives to maintain a disease-free status, there will often be some incentives for free-riding at the individual grower level. This can occur if growers fail to monitor or eradicate in an effort to reduce costs. Therefore, there is potential for market failure at both the prevention and eradication stages and thus a role for government intervention at both levels.

When a disease outbreak occurs, a public policy framework in the form of a benefit-cost test is usually employed to justify control and eradication measures. This often involves a mixture of public and private costs being incurred to avoid more wide-spread private costs if disease became established.

Governments have a range of policy mechanisms to use when addressing disease outbreaks. These include the use of public information and suasion methods to encourage appropriate action and voluntary compliance. Incentives for growers to voluntarily report, control and monitor outbreaks can also be enhanced with the use of financial instruments. Regulatory and control measures are an alternative or extensions to voluntary mechanisms, and have advantages in terms of response certainty and timelines.

However, the use of regulatory approaches to control may generate offsetting flow-on effects if they reduce growers' incentives to report and monitor outbreaks.

Therefore, there is a number of economic issues raised in disease control measures. From the perspective of society, it is important to determine whether the control and eradication measures are worth the cost of implementation. An economic analysis is considered appropriate for this purpose.

ECONOMIC ANALYSIS APPROACHES TO DISEASE OUTBREAKS

Broadly speaking, two approaches are used to estimate the economic cost of an incursion, namely impact modelling and economic surplus analysis. Impact modelling can take the form of an input-output (I-O) model or a computable general equilibrium (CGE) model.

An I-O model captures the supply and demand of goods and services in an economy (within an industry, region and economywide) in a particular period, as well as the interdependencies among the industries and associated primary factors of production.

By tracing these linkages between sectors, the model estimates the effect of an impact on the region's output, employment, income and imports, often expressed in the form of multipliers.

An I-O model can be used to capture the direct effects (output, income and employment) of a change in demand or supply as well as the indirect or flow-on impacts on its ancillary sectors (suppliers of inputs and process industries). It can also be used to measure the demand substitution effects, that is, consumers' shift of consumption to substitute commodities. Computable general equilibrium (CGE) models are constructed on the same data as the input-output models. A CGE model can provide information about the magnitude and sources of economic losses and the appropriate design of mitigation strategy (for example, eradication vs containment).

Both I-O and CGE models are able to capture the economy-wide impacts and inter-industry linkages, however, a CGE model has greater flexibility in terms of underlying assumptions and generates less biased estimates compared to I-O model.

Furthermore, CGE models have advantages over I-O models in that they can be used to estimate not only the economic impacts but also the distributional impacts. However, I-O models have advantages in that they help to understand the social and political impacts.

Elliston, Hinde and Yainshet assessed the economic impacts of an incursion of karnal bunt in a hypothetical situation in South-eastern Queensland using an input-output model. This study, however, failed to capture the changes in economic surplus.

The I-O model estimated the changes in output and employment, not the economic welfare, making it difficult to assess the overall gains or losses to society. Wittwer, McKirdy and Wilson estimate both the micro and macroeconomic effects of a hypothetical incursion of karnal bunt in wheat in the wheat belt of Western Australia using a CGE model. The analysis showed that quarantine restrictions in foreign markets dominated economic losses of the incursion of exotic plant diseases or pests.

Using the MONASH Multiregional Forecasting Model (MMRF), a version of CGE model, Dent *et al* estimated the likely impact of a foot-and-mouth disease (FMD) outbreak in Queensland and other regional economies. The study estimated the direct impact on 'at risk' industries and associated products as well as the flow-on effects to other industries and the economy as a whole.

The study found that an outbreak of FMD would impact not only the livestock farming and meat processing industries, but also other sectors in the economy (for example, tourism). The impact of the hypothetical FMD outbreak in Queensland (a major beef cattle state) was proportionately more severe than the national impact. In Queensland, the real GDP was projected to be $2,340 million below the base case, while employment was projected to be 33,900 jobs below the base case in Year 7 alone.

These losses are far greater than the projected Year 7 losses in the national livestock farming industry and the national livestock product (mainly meat processing) industry ($200 million and 5100 jobs). Although the estimated total control cost of FMD outbreak ($500 million) appears significant, however, it is considered to be minor when compared to the potential loss of the national economy.

Economic surplus analysis is used to estimate the net gains and losses of different impacts to society, and is normally preferred to impact modelling as a more appropriate measure of welfare changes.

Within this framework, a partial equilibrium approach can be used to estimate the net losses due to an incursion, and the distribution of such losses between producers and consumers, expressed as changes in producer and consumer surplus. The economic surplus approach can be performed at a case study level, where the net surplus can be estimated with the application of cost-benefit analysis (CBA).

With this method it is possible to demonstrate how a disease outbreak can reduce the welfare gains which might otherwise have been obtained from the industry in the absence of an incursion. The economic surplus method has been used to estimate the cost of weeds in annual winter cropping system in Australia, and to measure the benefits of the virus prevention programme in deciduous fruit trees in the US.

Jones *et al* estimated the economic costs of weeds and their distribution in Australian annual winter cropping systems using the economic surplus model.

The loss of economic surplus due to weed infestations and weed control expenditure was estimated at $1,133 million in 1999. The distribution of total economic surplus suggested that producers appropriate the bulk of the economic welfare gained from the control measures (about 95 per cent), while consumers gained to a small extent (about 5 per cent) from the reduced market prices resulting from increased production due to weed control. The economic surplus approach measured both direct and indirect costs, but non-monetary impacts were not considered.

Cembali *et al* estimated net economic benefits of a virus prevention programme for apples, sweet cherries and clingstone peaches in the United States at the nursery, grower and consumer levels as changes in consumer and producer surpluses. Empirical estimates of the expected yield losses at both nursery and grower levels were made using the method of avoided losses.

Avoided costs were used to estimate programme benefits in three economic sectors: nurseries (avoided change in producer surplus), producers (avoided change in consumer and producer surpluses), and consumers (avoided change in consumer surplus). The empirical estimates suggested that the total benefits for all three sectors were approximately $227.4 million a year, or more than 420 times the cost of the programme.

Paarlberg, Lee and Seitzinger measured the welfare effects of an FMD outbreak in the United States using the economic surplus method. They decomposed the effect by groups, including livestock producers with animals quarantined and slaughtered and producers not quarantined, using lost sales and producer surplus measures respectively. Similarly, consumer surplus was decomposed for consumers with and without changes of consumption behaviour (that is, whether or not consuming beef considering potential human health effects from an outbreak).

The researchers argued that decomposition of aggregate welfare (consumer or producer surpluses) for different groups could provide more accurate estimates of changes for the policy-decision, and help to design compensation provisions.

Rabbits impose economic costs on wool producers, governments, communities and the environment. Vere, Jones and Saunders provided an economic analysis of rabbit control measures in the Australian temperate pastures through integrating estimates of the costs of rabbits in the pasture systems and the benefits of implementing control practices from the introduction of rabbit haemorrhagic disease (RHD). The approach taken was to:

- Determine the changes in livestock production at the farm level due to supply shifts;
- Simulate these changes within a livestock industry model (that is, quantities, prices and elasticities);
- calculate the resulting economic welfare changes (that is, net benefits) and the relative benefits and costs of improved rabbit control over time.

The study showed that controlling rabbits in temperate pastures by RHD had the potential to generate substantial long-term economic benefits – the benefit-cost ratios were between 2.9:1 and 16.2:1 for a 25 per cent rabbit reduction and for a 50 per cent reduction the ratios were between 5.9:1 and 32.4:1. Zansler, Spreen and Muraro employed cost-benefit analysis to determine whether the citrus canker eradication programme in Florida could be a useful policy tool in combating the economic ramifications associated with the outbreak using the predicted values of the benefits and the costs associated with the intervention.

The results of the benefit-cost analysis suggested that benefits of the programme outweighed the costs.

This review of current literatures on plant or pest disease incursions demonstrates that both economic surplus and CBA cannot sufficiently capture the indirect or flow-on effects of an impact. On the other hand, I-O or CGE model cannot take into account non-market impacts or estimate net welfare change. Both approaches have their strengths and limitations. In this paper, the approaches of economic surplus and CBA are used to indicate the economic efficiency of citrus canker outbreak control and eradication strategies in Queensland.

CASE OF CITRUS CANKER OUTBREAK

The citrus industry in Australia makes a substantial contribution to the national economy and the generation of employment. Major citrus fruits in Australia include oranges (navel and valencia), mandarins, lemons, limes and grapefruit.

In 2002-03, citrus production in Australia consisted of 81 per cent oranges (valencia 45 per cent and navel 36 per cent), 14 per cent mandarin, 4 per cent lemon and lime and 1 per cent grapefruit.

The industry produces around 830 thousand tones of fruit per year with a gross value of production of $426m in 2002-03. Citrus is one of the major horticultural exports in Australia, generating around $160m of annual export income and contributing $250m directly to value-added products.

Exports accounted for around 20.6 per cent of total production in 2002-03. There was a record export of citrus by value of $201m in 2002-03. The focus of both the production and export of citrus is on fresh fruits and processed juice products.

The citrus industry is Australia's largest fresh fruit exporting industry. Commercial citrus production in Queensland is localised in the Central Burnett and Central Highland regions; out of 3,000 growers (cultivating 32,000 hectares of land) in Australia, about 300 are located in these regions.

Seventy five per cent of Queensland's citrus is produced in the Gayndah- Mundubbera (Central Burnett) region and the Emerald Shire (Central Highland region); the remainder is grown in Bundaberg, Capricorn Coast Region, Wide Bay, Mareeba (Far North), Sunshine Coast and some other small patches of land in the State.

The citrus industry generates about $80m annually to the Queensland economy and supports several thousand jobs, particularly for seasonal workers. The majority of fruit is transported to southern markets and some is exported.

Citrus canker is a highly contagious disease that causes defoliation and dieback, severely blemished fruit, weakens trees and eventually reduces production, although it is not harmful to humans. The disease was first discovered in Queensland in early June 2004 on a 1,200 hectare (ha) orchard with 220,000 citrus trees (Evergreen Farm) near Emerald in Central Highlands region. Immediately after the detection of the incursion, a Pest Quarantine Area (PQA) was declared for the Shires of Bauhinia, Emerald and Peak Downs, surrounding the infested orchard. As well, an interstate ban on the movement of *Rutacae* (the plant family to which citrus belongs) fruits was imposed.

The trade ban was also imposed on the Gayndah/ Mundubbera region due to its proximity to Emerald citrus growing region and the possibility that disease might spread through the movement of plant material and seasonal workers. Growers were restricted from trading fruit or plant material within Queensland or interstate.

The interstate trade of citrus for fruits grown outside of the Central Highlands and the Gayndah/ Mundubbera region resumed on the 22 July, while for citrus growers in the Gayndah/Mundubbera region trade resumed on the 13 January 2005.

Furthermore, all Australian citrus growers were required to undergo stricter quarantine checks before exporting fruit to New Zealand pending quarantine surveys to establish state, area or property freedom from citrus canker.

While the European Union stopped importing citrus from Australia until a canker-free status was established for the whole country, export to other markets, particularly canker infected markets in Asia, did not face any restriction. Citrus canker was widespread on the first affected property. Under the National Citrus Industry Biosecurity Plan (NCIBC) introduced in April 2004, a 'cookie cutter' approach involving destruction of trees within 600 metres of an infected area is considered to be effective control. By late 2004, all the 220,000 trees on the first affected property were taken out with successive 'cookie cuts'.

A second incursion within the Emerald irrigation area was reported on October 5, 2004 on Queensland's largest citrus orchard, known as 2PH Farms, having 326,000 fruit trees. This orchard is only 7 km away from the first infested farm and subsequent DNA testing confirmed that this latter outbreak was linked to the first one. By July 2005, three orchards in Emerald were affected with six outbreaks and over half a million citrus trees had been destroyed.

Some of the revenue and expenditure impacts on the Emerald region can be estimated as follows:

- Annual loss of revenue to growers from tree removal as of July 2005 was $21. 3m/year (500,000 trees at $42. 6 estimated annual production/tree);
- Annual loss of revenue to growers if all citrus trees removed from Emerald would be $31. 95m/year (750,000 trees at $42.6/tree);
- Surveillance and control cost allocation by government was $13m (perhaps more than 50 per cent spent in the Emerald region);
- Several hundred full-time jobs and more than one thousand part-time jobs are anticipated to have been lost.

With disease outbreak in livestock, compensation is automatically paid.

However, this was not the case with a citrus canker outbreak as no cost sharing deed for emergency pest management was in place during the Emerald incursion. The Queensland citrus industry has been affected by the loss (short to midterm) of some overseas markets and if domestic consumers reduce demands for citrus fruits. Such losses will flow the temporary ban on interstate trade.

Losses may also extend to domestic markets, especially through to other parts of the economy, such as transport, wholesale and retail sectors.

On the eradication side, the citrus canker outbreak has generated higher control costs for private industry, as well as considerable costs for eradication and control by the regulatory authorities. This raises questions as to the impact on Queensland's citrus industry and society as a whole, and whether public expenditure on control and eradication was an efficient use of funds.

THE PRESSLER AND THE NATIONAL CITRUS INDUSTRY BIOSECURITY PLANS

A debate has taken place about the most efficient way to perform control and eradication measures in the region. Under the NCIBP, all trees within 600 meters of an infection are removed, and then monitoring continues. After the first disease detection on the second farm, the owner, John Pressler, put forward a proposal to destroy all the citrus plants in Central Highlands region around Emerald.

The core of this proposal, referred to in the media as the Pressler Plan, was for growers to destroy all remaining trees with financial support from the Government. This would involve funding for the growers of $16m ($50 payment for destruction per tree).

This would reduce any chance of canker spreading and then allow the area to re-plant after two years. The rationale behind the Pressler Plan was that if the current eradication programme failed to eliminate the disease completely from the region, on-going outbreaks would lock growers out of markets for long periods of time.

If implemented, the Pressler Plan would provide growers with certainty about their re-establishment and market access. After the destruction, the growers will be required to fallow the land for the next two years, replant and wait for the new trees to bear their first crop.

An orchard will return to its production in five to six years time. Considering Emerald's isolation from the other growing regions in the

State and the high probability of returning to production, this proposal received support from the growers in the region. However, the government rejected the Pressler Plan, preferring to continue with the response strategy under the nationally approved response plan.

It was argued that the response plan would be effective and that the risk of continued outbreaks was very low. Subsequent to the rejection of the Pressler Plan, two further outbreaks were discovered at the 2PH farm, and another outbreak discovered at a third farm in the region. Finally, the National Management Group in June 2005 supported the removal of all citrus trees, including domestic and wild native trees in the Emerald PQA, but without any provision of compensation or voluntary agreement with growers. The effectiveness of the NCIBP in eradicating the disease is a key issue.

Citrus canker is still endemic in Florida (USA) despite applying the 'cookie cutter' approach for more than a decade in response to citrus canker outbreaks. Grower and industry concern at Emerald was that under the 'cookie cutter' approach all citrus trees in the region might eventually face destruction as further infections were detected.

EVALUATING THE RESPONSE STRATEGIES

An economic analysis of quarantine and control measures can take two basic forms. The first form of analysis is to identify the net benefits of control measures. The net costs of control can be estimated as the changes in producer and consumer surpluses, plus the public expenditure committed to the programme.

The benefits of the control measure include the avoidance of further public expenditure on controls, the protection of producer and consumer surpluses, and the avoidance of public expenditure on industry and regional community support. Given the size of the citrus industry in Australia, the potential public expenditure involved in controlling a major outbreak, and the vulnerability of many regional communities to a downturn in the citrus industry, it is likely that there are major net benefits associated with controlling the disease.

The averted cost framework can be used to evaluate this issue. For the Pressler Plan to have provided net benefits, the estimated $16m cost (destruction cost of about 326,000 remaining trees in the Emerald PQA) should have been lower than the cost of potential impacts multiplied by the likelihood of outbreaks occurring.

The total cost of a major outbreak is difficult to estimate, but some broad assumptions can be used to help with the analysis. If a major outbreak occurred across the citrus industry, markets would close and there would be substantial control costs.

If producer surplus is estimated at 10 per cent of current industry production ($426m/year), and production is affected for at least five years to return to production, the discounted present value of producer surplus, at an 8 per cent discount rate, may be estimated at $170m. To this should be added potential control costs, as well as additional welfare payments associated with support for unemployment and impacts on regional communities.

Given that a potential control cost of $13m has been allocated by the government for the Emerald outbreaks, a potential control cost of $40m may be appropriate for a wider national outbreak.

If the citrus industry is closed down in Australia for 5 years, over $2 billion in direct revenue would be taken out of mainly rural communities.

It is not unrealistic to expect that the public support would be needed to account for the total impact. At a 10 per cent support level, total costs (loss of producer surplus) would be $200m. To summarise, an eradication measure for citrus canker has net benefits because it averts three main types of potential costs in the Australian situation:

- Losses in producer surplus (effectively the present value of industry profits in Australia);
- Avoided expenditure on control measures for major outbreaks; and
- Avoided expenditure on industry reconstruction and community support.

Using the rough estimates provided above, the economic benefits of averting a national outbreak of citrus canker are $410m, while the public cost of eradication was estimated to be $13m. There are clear benefits in pursuing eradication measures.

The second form of analysis is to identify the level of control that yields the highest net benefits. An example of this approach can be demonstrated by evaluating the net benefits of the Pressler Plan compared to the NCIBP.

Here, the difference in the costs of control measures between the plans is compared. Both the tangible and intangible costs and benefits between the Pressler Plan and current NCIBP are summarised in Table.

Table: Additional Benefits and Costs of the Pressler Plan over the National Citrus Biosecurity Plan

Benefit	*Cost*
Reduced chance of further outbreaks and hence certainty of returning to faming by growers	Payments to growers for destruction of tree ($50/tree)
Reduced inspection costs by goverment over longer term	Replacement costs to replant trees
Increased certainty about eradication and hence avoided loss of domestic and international (especially USA and EU) markets	Loss of production & sale for 5 years (net with other crop return)
Avoided public costs on control of any further outbreaks	Potential impacts on other industries (e.g. grapes and melons) if growers diversify into other crops
Private benefits of replanting with better fruit stock	Impacts on households of removing all domestic citrus trees

The key issue is whether the reduced risk of eradication represented by the Pressler Plan would have generated sufficient benefits to justify the increased costs. Jetter, Civerolo and Sumner reporting for citrus canker infestations in California, estimated an eradication boundary of 1,900 feet (579 meters) resulted in a 95 per cent chance of eradication. The probability of eradication increased to 99 per cent if the boundary was increased to 3,000 feet (914 meters).

There have been multiple outbreaks of citrus canker in the United States as the 1,900 feet 'cookie-cutter' approach to eradication has not been fully successful in controlling the disease, indicating that a 95 per cent probability may be overestimated.

If the probability of a subsequent outbreak under the NCIBP is only 5 per cent, then the probability of several successive outbreaks is extremely low. Given that the latter has occurred at Emerald, it appears that the real probability of an outbreak is higher than 5 per cent. If the current NCIBP is assumed to have a 95 per cent chance of eradication and the Pressler Plan assumed to have a 100 per cent chance, the key issue is whether a 5 per cent increase in the probability of eradication would have been justified in comparison to the change in net costs. The key additional cost involved in the Pressler Plan was $16m to be paid to growers as partial compensation for the removal of the remaining trees.

The total cost of a major national outbreak has been assessed at $410m. This amount should then be multiplied by the potential risk of occurrence to determine the appropriate avoided cost to consider. At 10 per cent, 5 per cent and 2.5 per cent levels for risk of occurrence, these expected benefits are $41m, $20.5m and $10.25m respectively.

Comparing this to the expected cost of $16m suggests that the Pressler Plan would have delivered net benefits at the 10 per cent and 5 per cent risk levels, but not at the 2.5 per cent risk level. This indicates that the Pressler Plan was justified on an economic basis when the mitigation strategy has only a 95 per cent probability of eradication.

With the 'cookie cutter' approach, the risk of occurrence still remains. Experience in Florida showed that as the boundary of the eradication zone increased, the eradication costs also increased.

Spreen and Zansler estimated the economic costs to Florida citrus growers in two scenarios: production costs associated with the eradication of citrus trees due to a canker infestation, and production costs associated with living with citrus canker. A hypothetical canker-free orchard was modelled for the comparison. Results indicated that the estimated economic loss to citrus growers in an endemic citrus canker situation exceeded the estimated economic loss to citrus growers under eradication.

Studies in Florida also showed that 90 per cent of the citrus canker infections occurred within the1,200 feet buffer, but wind-borne spread could carry viable bacteria as far as 3,900 feet. Using Argentina as an example for a Florida citrus canker control programme, Muraro, Roka and Spreen concluded that compared to 'living with canker', eradication was a worthy choice if the disease was identified early and the action was taken swiftly and decisively.

EXOTIC PLANT PESTS AND DISEASES

The threat to biosecurity from exotic plant pests and diseases is a major one for Australia, where a single incursion can generate large commercial, economic and social costs. Evidence of this comes from the recent outbreak of citrus canker in orchards at Emerald in Central Queensland. It is not only local growers who bear costs, as a range of other stakeholders (including taxpayers) are potentially affected. To design an appropriate response strategy, it is important to be able to analyse the impacts and costs in economic terms, and provide a comparison of the potential benefits and costs of different response strategies.

There is controversy concerning the efficacy and economic efficiency of the appropriate response mechanisms to citrus canker outbreaks in Emerald. The public funding allocation for eradication is estimated at $13m; there will also be a range of private costs involved in control and eradication.

The averted costs of the eradication measures are very large, indicating there are substantial net benefits associated with complete destruction of citrus trees in the region. Questions can be raised about the effectiveness of the NCIBP in two main areas.

The first is that plant growers are not compensated for tree destruction in disease control measures. This contrasts to the compensation provided to livestock producers for animals destroyed to control disease (for example, brucellosis outbreaks in the cattle industry).

The lack of compensation measures means that growers may face large individual incentives not to report disease outbreaks. Some form of compensation appears to be justified on equity and efficiency grounds, and given the distribution of the benefits of control measures, should be funded by both government and industry. The second issue about the effectiveness of the NCIBP is that the likelihood of success was overestimated. A basic economic analysis has shown that the Pressler Plan would have been a more efficient control measure if the probability of the 'cookie cutter' approach being successful was 95 per cent or lower.

Given that this approach failed, not once, but several times, to control the disease at Emerald, the probability of success of the 'cookie cutter' approach would appear to be much lower than 95 per cent. The economic analysis in this paper indicates that the government should have accepted the Pressler Plan. These results suggest that the NCIBP is flawed.

The control of further disease outbreaks is likely to be enhanced:

- If there are better incentives for individual growers to comply with the plan;
- If more accurate assessment of the probability of success of the 'cookie cutter' approach is provided;
- If more widespread control measures like the Pressler Plan can be considered as the most cost-effective control strategies from the perspective of society.

The use of economic analysis to identify the potential benefits and costs of different levels and types of control measures is recommended as part of a review of the control and eradication strategy.

This analysis demonstrated that the total net benefit of a complete destruction of the Emerald citrus trees outweigh the costs to control and industry losses under the scenario of at least 95 per cent probability of eradication. To strengthen on-farm biosecurity measures, growers infected with canker can be compensated to offset the loss with reporting the disease and being placed under quarantine restrictions.

